University
LECTURE
Series

Volume 46

Topology of Tiling Spaces

Lorenzo Sadun

American Mathematical Society
Providence, Rhode Island

2000 *Mathematics Subject Classification.* Primary 52C22, 55–02, 52C23, 55N99, 55N05.

Figure 1.4 (bottom), from Penrose, Roger, "Mathematics of the Impossible", in Gregory, R. L., et al. (editors), *The Artful Eye*, Oxford University Press (1995), p. 326, Figure 16.2, is used with the kind permission of Oxford University Press and Roger Penrose; Figure 7.3 is used with the kind permission of Dr. Natalie Priebe Frank.

For additional information and updates on this book, visit
www.ams.org/bookpages/ulect-46

Library of Congress Cataloging-in-Publication Data

Sadun, Lorenzo Adlai.
    Topology of tiling spaces / Lorenzo Sadun.
       p. cm. — (University lecture series : v. 46)
    Includes bibliographical references.
    ISBN 978-0-8218-4727-5 (alk. paper)
    1. Tiling spaces. 2. Aperiodic tilings. 3. Topology. 4. Tiling (Mathematics) I. Title.

QA611.3.S23   2008
516′.132—dc22
                                         2008029389

For my children: Rina, Allan and Jonathan.

May their lives be like an aperiodic tiling —
infinitely varied, but with everything fitting perfectly.

Contents

Preface

This book is intended as a very personal introduction to the topology of tiling spaces, with a target audience of graduate students who wish to learn about the interface of topology with aperiodic order. It isn't a comprehensive and cross-referenced tome about everything having to do with tilings. That would be too big, too hard to read, and way too hard to write! Instead, I have tried to lay out the subject as I see it, in a linear manner, with emphasis on those developments that I find to be the most interesting. In other words, this book is about how I think about tilings, and what inspires me to keep working in the area. My hope is that it will also inspire you.

"Interesting" is a subjective term, of course. Many subjects that others consider to be central are not covered here. For instance, you will find little about cut-and-project tilings in this book, despite the mass of work that has been done on them. I mean no disrespect to the practitioners of that field! I just can't do that subject justice, and am happy to leave its exposition to people who know it far better than I. (For cut-and-project tilings, I particularly recommend Bob Moody's review article [**Moo**] and, for the ambitious, the comprehensive monograph by Forest, Hunton and Kellendonk [**FHK**].)

By contrast, I love inverse limit structures, tiling cohomology, substitution tilings and the role of rotations. I love pattern-equivariant cohomology. I love tilings that don't have finite local complexity. In this book you'll see them all, in considerable detail.

Modern tiling theory developed from four very different directions. One direction was from logic. In the 1960s, Hao Wang and his students posed a variety of problems in terms of square tiles with marked edges (aka Wang tiles). Given a set of such tiles, can you determine whether it's possible to tile the plane in a way that the edges of adjacent tiles match? Wang's student Berger [**Ber**] proved the answer to be "no" in general, and in the process produced a set of tiles that would tile the plane but only nonperiodically. Berger's example was extremely complicated, but people quickly produced simpler examples.

The second ingredient was a good example for study and wonder. Roger Penrose produced a set of aperiodic tiles in the mid-70s that sparked intense interest and brought aperiodic tilings into popular culture. The Penrose tilings [**Pen**] aren't just mathematically interesting – they're pretty! They also demonstrate a remarkable rotational symmetry; every pattern that appears somewhere in the tiling also appears rotated by 36 degrees, and with the same frequency. This "statistical symmetry" contradicts the long-held belief that only rotations by 60, 90, 120 and 180 degrees can appear in highly ordered structures.

The third ingredient came from physics, or you might say from materials science. In 1982, Shechtman and coworkers [**SBGC**] discovered a new class of solid, neither crystal nor amorphous, called *quasicrystals*. Quasicrystals have sharp diffraction patterns, long thought to be the hallmark of a periodic crystal, but some of these patterns have 8- or 10-fold rotational symmetry. It didn't take people long to realize that quasicrystals are modeled well by aperiodic tilings, and in particular by 3-dimensional versions of the Penrose tiling and by several other cut-and-project tilings!

The fourth ingredient came from ergodic theory and dynamical systems, where substitution sequences had long been a subject of interest. Some of the simplest substitutions, like the Thue-Morse substitution, were defined over 100 years ago. However, it was only in the 1980s that people went from substitution subshifts to substitution tilings in one dimension, and from there to higher-dimensional substitution tilings. It didn't hurt that the Penrose tiling could be realized in this way. Soon people discovered other interesting geometric "rep-tiles" and computed properties of the resulting tilings.

These trends came together in the 1990s. Tilings, including the Penrose tiling, were used to model quasicrystals. These tilings were in turn generated in a number of ways, including local matching rules, cut-and-project methods, and substitutions. The tilings were then studied as dynamical systems, and their dynamical properties were related to physical properties of the quasicrystals that they model.

Suppose you had a quasicrystal that was modeled by an aperiodic tiling. A physicist might ask the following questions about the quasicrystal.

- P1. What is the x-ray diffraction pattern of the material? This is equivalent to the Fourier transform of the autocorrelation function of the positions of the atoms. Sharp peaks are the hallmark of ordered materials, such as crystals and quasicrystals.
- P2. What are the possible energy levels of electrons in the material? The locations of the atoms determine a quasiperiodic potential, and the spectrum of the corresponding Schrödinger Hamiltonian has infinitely many gaps. What are the energies of these gaps, and what is the density of states corresponding to each gap?
- P3. Can you really tell the internal structure of the material from diffraction data? What deformations (either local or non-local) of the molecular structure are consistent with the combinatorics of the molecular bonds? Which of these are detectable from diffraction data?

Although the physicist is interested in a single quasicrystal (or a single tiling), mathematicians like to define *spaces*. From the tiling T we construct a space Ω_T of tilings that have the same properties of T. If T has desirable properties (like finite local complexity, repetitivity, and well-defined patch frequencies), then Ω_T has corresponding properties (compactness, minimality as a dynamical system, and unique ergodicity), and we can ask the following mathematical questions:

- M1. What is the topology of Ω_T? What does the neighborhood of a point of Ω_T look like? What are the (Čech) cohomology groups of Ω_T?
- M2. There is a natural action of the group $\mathbb{R}^d$ of translations on Ω_T. This makes Ω_T into a dynamical system, with d commuting flows. What are the ergodic measures on Ω_T? For each such measure, what is the spectrum of the generator of translations (think: momentum operator) on $L^2(\Omega_T)$? This is called the *dynamical spectrum* of Ω_T.
- M3. From the action of the translation group on Ω_T, one can construct a C^* algebra. What is the K-theory of this C^*-algebra?

Remarkably, each math question about Ω_T answers a physics question about a material modeled on T. M1 answers P3, M2 answers P1, and M3 (in large part) answers P2. Far from being a pointless mathematical abstraction, tiling spaces are important!

This book is the story of the first mathematical question, and the answers we have gleaned so far. In chapter 1 we consider a variety of interesting tilings and the construction of the corresponding tiling spaces. In chapter 2 we explore the local structure of Ω_T and its realization as an inverse limit space. In chapter 3 we introduce the Čech cohomology of Ω_T and show how the answer to the third physics question is tied to the first Čech cohomology. In chapter 4 we study the rotational properties of tilings — what made quasicrystals and the Penrose tiling so amazing in the first place! In chapter 5 we introduce "pattern-equivariant cohomology", a beautiful realization of tiling cohomology in terms of properties of an individual tiling. In this way we come full circle, from tilings to tiling spaces and back to individual tilings.

The material in the first four chapters is basically set in place, as is some of the material of chapter 5. Chapters 6 and 7, however, are cutting-edge research. The reader may have some difficulty with these chapters, both because the concepts aren't as neatly prescribed as the earlier topics, and because the calculations require more advanced algebraic topology.

Chapter 6 is devoted to "tricks of the trade", recently developed calculational techniques that are powerful but are *not* generally known. Chapter 7 is about tilings without the simplifying assumption of finite local complexity. Until recently, such tilings were thought to be beyond our understanding, but that is rapidly changing.

Exercises for the reader are embedded in the text. I strongly recommend that you work these out carefully. If you run out of patience, you can look up the answers to most of these exercises in the appendix.

The first five chapters are largely based on a series of lectures that I gave at the 2005 Summer School in Aperiodic Order at the University of Victoria. I am grateful to the organizers and participants in this summer school for their enthusiasm and their moral support. I am also especially grateful to Margaret Combs for her TEXnical and artistic assistance. This book is partly based upon work supported by the National Science Foundation under Grant No. 0701055.

CHAPTER 1

Basic notions

1.1. Tilings

A tiling is a subdivision of the plane (or more generally, of $\mathbb{R}^d$) into pieces called tiles. Specifically, we have a set of tiles t_i that intersect only on their boundaries, and whose union is the entire plane. Technically, the tiling is that *set* of tiles. Besides their geometric shapes, tiles may carry labels – think of them as the color of the tiles. A *patch* of a tiling is a finite subset of the tiles in a tiling. If A is a bounded subset of $\mathbb{R}^d$, then $[A]$ denotes the patch consisting of all tiles that intersect A. Sometimes we will write $[A]_T$, to emphasize that we are talking about a patch of the tiling T. If T is a tiling and $x \in \mathbb{R}^d$, then $T - x$ is the same set of tiles shifted over by the vector $-x$. This is equivalent to moving the origin by $+x$, and generally results in a different tiling. In T there may be a tile near the point x, while in $T - x$ that tile appears near the origin. Moving a tiling around like this is one of the most important operations that we will consider.

Although tiles are usually polygons, sometimes they take shapes that are artistically more interesting. Several interesting tilings appear on the following pages.

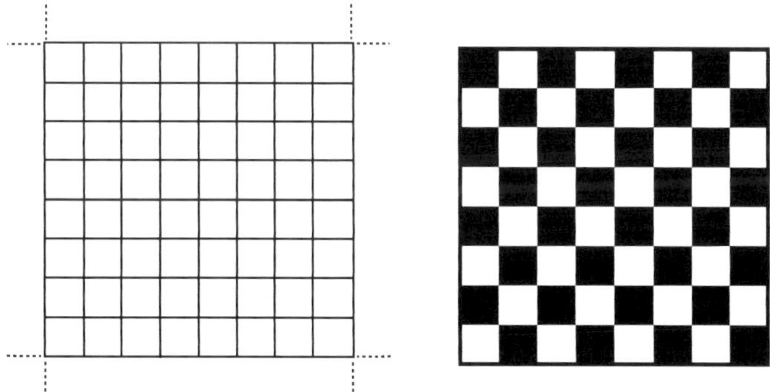

FIGURE 1.1. Basic square tiling and basic checkerboard.

The basic square tiling, shown in Figure 1.1, is frequently found in kitchens. There is only one kind of tile, and it is repeated over and over again. Not very interesting. A checkerboard is only slightly more complicated. There are now two kinds of tiles. Both are unit squares, but one is called "black" and the other is called "white". Translating this tiling to the left by 2 units, up by 2 units, or diagonally by $\sqrt{2}$ units gives the exact same tiling, and we say that this tiling is "periodic". As we shall see, periodic tilings have very little complexity; we will concentrate on tilings that are not periodic.

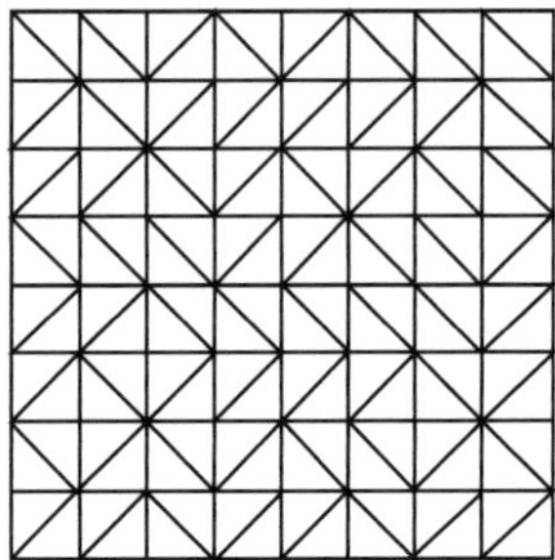

FIGURE 1.2. Random tiling of $\mathbb{R}^2$ by triangles.

Suppose we take a basic square tiling and cut each tile in half along a diagonal, as in Figure 1.2. We flip a coin to decide whether to cut from the bottom left corner to the top right corner, or from the top left corner to the bottom right corner. With probability 1, the result will not be periodic. Such a tiling is extremely complicated, but it has little structure. Knowing what the tiling is like in one region says nothing about what it is like in another region (beyond the fact that triangles fit back-to-back into squares that repeat periodically). While the square tiling has too much structure and not enough complexity, the random tiling has too much complexity and not enough structure.

We are mostly interested in something in between. We want tilings that have long-range order, but are not periodic. Knowing what the tiling looks like at one point should not completely determine the tiling far away, but it should say *something* about the tiling far away. As we shall see, there are many tilings that exhibit such "aperiodic order".

One such tiling is the "chair" tiling of Figure 1.3. There are four kinds of tiles, each consisting of a 2×2 square with one 1×1 corner removed. Each tile is part of a group of four tiles that form a larger chair-shaped region, a group of 16 that forms an even larger chair-shaped region, and so on. This is an example of a "substitution tiling". Most of the examples in this book will be substitution tilings.

The Penrose tiling [**Pen**] of Figure 1.4 is known for its 10-fold rotational symmetry. Any pattern that is seen in this tiling is also seen, with the same frequency, rotated by any multiple of 36 degrees. The Penrose tiling comes in several equivalent forms. The most famous is the "kites and darts" version. One can break the kites and darts into triangles. One can also bend the boundaries of the kites and darts into chickens. In the triangle version, there are four kinds of triangles, each of which can appear in 10 different orientations, for a total of 40 species of tiles.

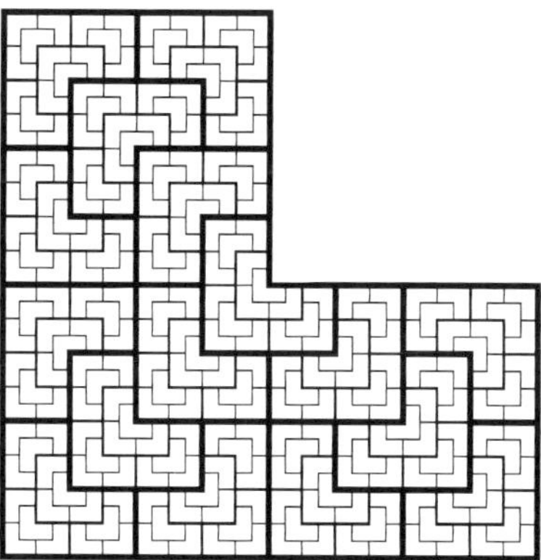

FIGURE 1.3. A patch of a chair tiling

The pinwheel tiling [**Rad**] of Figure 1.5 has two kinds of tiles: a 1-2-$\sqrt{5}$ right triangle and its mirror image. However, each kind appears in an infinite number of orientations! Like the Penrose tiling, this tiling has a statistical rotational symmetry. Any pattern that appears in the tiling also appears rotated, and the rotations are uniformly distributed in $SO(2)$.

So far our examples have all been tilings of the plane. However, the definition of tiling makes sense in arbitrarily many dimensions. What's more, many of the interesting phenomena already appear in one dimension! Figure 1.6 shows several interesting 1-dimensional tilings.

The first tiling, which we call "one black tile", has two kinds of tiles, both of length 1. One tile is black (say, on the interval $[0,1]$), and all other tiles are white. This tiling is not periodic, but it is still highly ordered. The "half and half" tiling has white tiles to the left of the origin and black tiles to the right. Like the "one black tile" tiling, it is not periodic, but is far from random. The Thue-Morse and Fibonacci tilings have a hierarchical structure, much like the chair tiling in two dimensions. In the Thue-Morse tiling, tiles group into pairs (either ab or ba), which group into collections of four (either abba or baab), which group into collections of eight, etc.

Our definition of tiling is unnecessarily broad, as it allows truly bizarre arrangements of truly bizarre shapes. Most of the time we will study something much simpler:

DEFINITION. *A simple* tiling of $\mathbb{R}^d$ *is a tiling in which*

(1) *There are only a finite number of tile types, up to translation. Put another way, there exists a finite collection of* prototiles p_i *such that each tile is a translated copy of one of the* p_i.

(2) *Each tile is a polytope. In one dimension, that means an interval. In 2 dimensions, it means a polygon (not necessarily convex). In three dimensions, it means a polyhedron.*

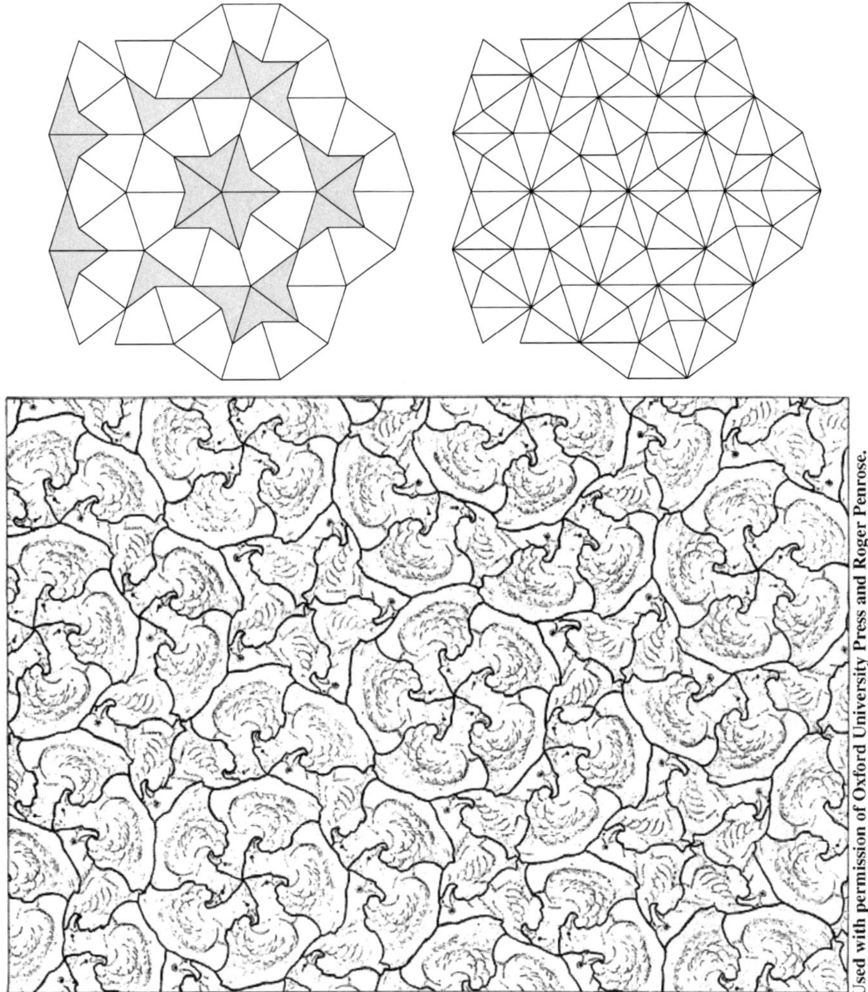

FIGURE 1.4. Penrose kites and darts, triangles, and chickens

(3) *Tiles meet full-edge to full-edge. An edge of one tile cannot overlap
with an edge of a neighboring tile. (In 3 dimensions, a face cannot
overlap with a face of a neighboring tile.)*

The pinwheel tiling is not "simple", because it violates the first hypothesis. There are only two kinds of tiles, up to rigid motion, but they appear
in infinitely many orientations, so there are an infinite number of tile types
up to translation. The first hypothesis can be relaxed to allow pinwheel-like
tilings, and we will consider such tilings in Chapter 4.

The Penrose chickens violate the second hypothesis. However, there is a
simple technique for converting tilings whose tiles are wild shapes into tilings
whose shapes are polygons. Pick a representative point for each tile (say, at
the center of mass of each chicken). Let p_i be the point corresponding to tile
t_i. Consider the set of points in the plane that are closer or equal in distance

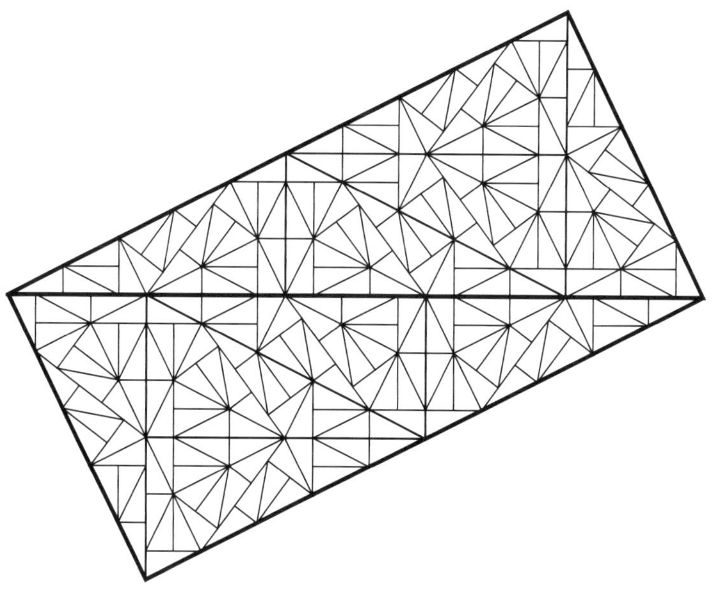

FIGURE 1.5. Patch of a pinwheel tiling

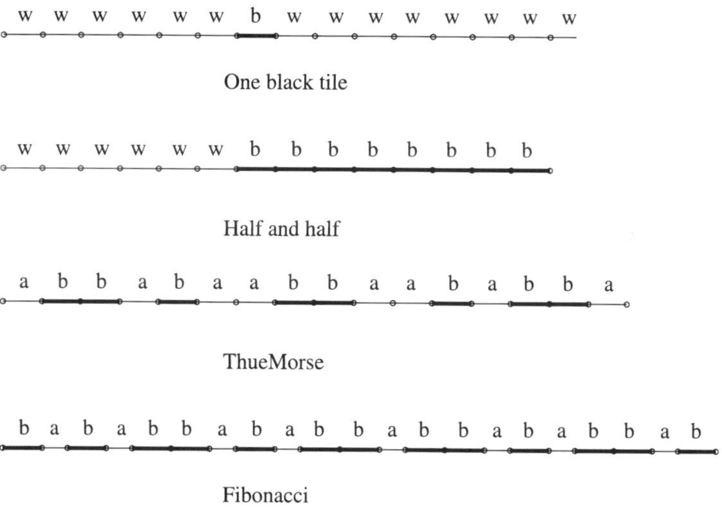

FIGURE 1.6. Patches of several one-dimensional tilings

to p_i than to any other p_j. This is called the *Voronoi cell* associated to p_i. The Voronoi cells are then polygons that tile the plane.

Finally, if we think of a chair tile as having six edges, then the chair tiling violates the third hypothesis. However, we can avoid this problem by adding two additional vertices, as in Figure 1.7, and thinking of each tile as a degenerate octagon. This trick is frequently used to convert tilings that don't seem to satisfy the third hypothesis into tilings that do.

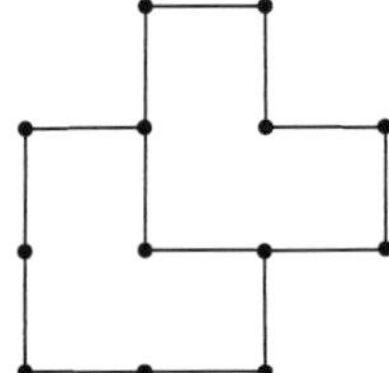

FIGURE 1.7. Two chair tiles. Do they meet full-edge to full-edge?

1.2. Tiling spaces

Now that we know what tilings are, where's the topology? The plane is contractible, as is $\mathbb{R}^d$. There's not much topology there! Instead, from each tiling T we will construct a space Ω_T of tilings and study the topology of Ω_T. As noted in the preface, mathematical properties of Ω_T are closely related to physical properties of materials modeled on T.

The first step is to define a metric on tilings. Given two tilings T and T' of the same $\mathbb{R}^d$, we say that T and T' are ϵ-close if they agree on a ball of radius $1/\epsilon$ around the origin, up to a translation of size ϵ or less. More precisely, let $R(T, T')$ be the supremum of all radii r such that there exist vectors x and y with $|x| < 1/2r$ and $|y| < 1/2r$ and such that $T - x$ and $T' - y$ agree on B_r, the ball of radius r around the origin. Then the distance $d(T, T')$ between two tilings is defined to be the smaller of 1 and $1/R(T, T')$.

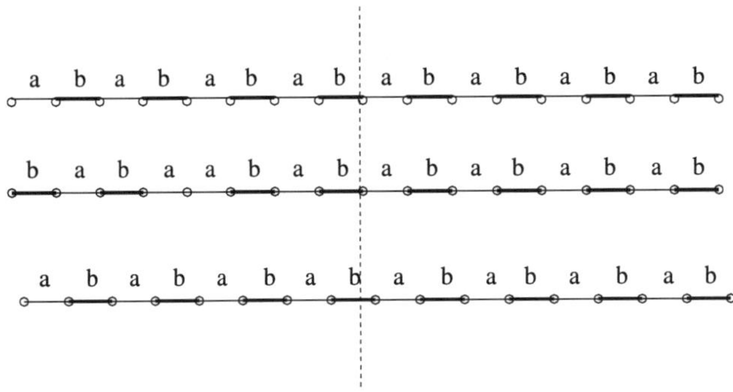

FIGURE 1.8. Three one-dimensional tilings. The first two are aligned but have different sequences, while the last is a translate of the first. The vertical line shows the location of the origin.

For instance, consider a tiling T and its translate $T - x$. If $|x|$ is small, then $d(T, T-x) = |x|$. However, tilings that are close to T are not necessarily translates of T. If a tiling T' is the same as T on B_r, but has a different tile at a distance r from the origin, then $d(T, T') = 1/r$.

In Figure 1.8, the first two tilings have distance at most 1/4 since they agree out to distance 4 from the origin. (The actual distance is slightly less than 1/4, since translating both tilings a small distance to the left will cause them to agree on a slightly larger ball.) The first and third have distance 1/4, since they agree everywhere, up to a translation by 1/4. The second and third have distance at most 8/31, since after translating the first to the right by $1/8 < 4/31$ and the second to the left by 1/8, they agree on a ball of radius 31/8 around the origin.

DEFINITION. *The* orbit *of a tiling T is the set $\mathcal{O}(T) = \{T - x | x \in \mathbb{R}^d\}$ of translates of T.*

DEFINITION. *A* tiling space *is a set of tilings that is (1) closed under translation, and (2) complete in the tiling metric. That is, if $T \in \Omega$ then $T - x \in \Omega$, and every Cauchy sequence of tilings in Ω has a limit in Ω.*

DEFINITION. *The* hull Ω_T *of a tiling T is the closure of $\mathcal{O}(T)$. This is sometimes called the "orbit closure" of T.*

The hull Ω_T should be viewed as the set of tilings that locally look like translates of T. In particular, a tiling T' is in Ω_T if and only every patch of T' is found in a translate of T. (Equivalently, a translate of every patch of T' is found in T.)

To see this, suppose that P is a patch in T', located somewhere in B_r for some $r > 0$. If T' is in the hull of T, then there are translates of T that agree with T' on arbitrarily large balls around the origin, hence that contain P. Conversely, let $P_r = [B_r]$ in T'. If every patch of T' is found in a translate of T, then there exist translates T_r of T that contain P_r. But then $T' = \lim_{r \to \infty} T_r$ is in the orbit closure of T.

Examples of Hulls. To build our intuition, let's look at the hulls of some simple 1-dimensional tilings. Let T_0 be a periodic tiling with just one kind of tile: a white tile of length one. Since $T_0 - 1 = T_0$, the orbit of T is topologically the circle $\mathbb{R}/\mathbb{Z}$. This is already a complete metric space, so Ω_{T_0} is just a circle.

Next consider the "one black tile" tiling T_1. Every tiling with one black tile is in $\mathcal{O}(T_1)$, and hence in Ω_{T_1}. However, tilings with *no* black tiles are also in Ω_{T_1}, since every patch of such a tiling can be found in T_1, both sufficiently far to the right of the origin and sufficiently far to the left. Ω_{T_1} consists of two path-components: the circle Ω_{T_0}, and the line $\mathcal{O}(T_1)$, with both ends of the line asymptotically approaching the circle, as in Figure 1.9.

EXERCISE 1.1. *What is Ω_T when T is the half-and-half tiling? How many path components does it have?*

EXERCISE 1.2. *Let T be a 1-dimensional tiling, with the color of the tiles (black or white) decided by fair and independent coin flips. What is Ω_T? Your answer could in principle depend on T, since it's conceivable that your coin flips would always yield heads (T_0) or yield tails exactly once*

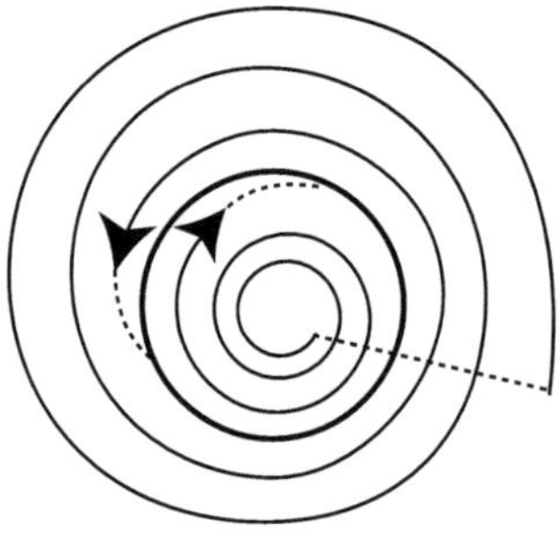

FIGURE 1.9. The hull of "one black tile"

(T_1). However, those are probability zero events. There is a space that Ω_T will equal with probability one. Find it.

THEOREM 1.1. *If T is a simple tiling, then Ω_T is compact.*

PROOF. We must show that every sequence in Ω_T has a convergent subsequence. Since there are only a finite number of tile types, and there are only a finite number of ways in which tiles can abut, for each r there are only a finite number of possible patches $[B_r]$, up to translation by a distance smaller than the diameter of the largest tile. Therefore, in any sequence of tilings in Ω_T, there is a subsequence that converges on B_r. Now apply the Cantor diagonalization trick. Of the subsequence that converges on B_1, pick a sub-subsequence that converges on B_2, a subsequence of that that converges on B_3, and so on. Take the first element of the sequence that converges on B_1, the second element of the sequence that converges on B_2, etc. This sequence converges on every bounded set, and so forms a Cauchy sequence in the tiling metric. Since the tiling space is complete, this sequence has a limit in Ω_T. $\square$

1.3. Equivalence

In topology, we are always classifying spaces up to equivalence. But what does it mean to say that two tiling tiling spaces are equivalent? There are several different notions, the weakest of which is homeomorphism.

DEFINITION. *A homeomorphism between (simple) tiling spaces is a continuous map $f : \Omega_T \to \Omega_{T'}$ that is 1-1 and onto. Since Ω_T is compact, f^{-1} is automatically continuous, so this agrees with the usual topological definition of homeomorphism.*

Homeomorphisms preserve topology, but little else.

DEFINITION. *A factor map between tiling spaces is a map that commutes with the action of the translation group. A topological conjugacy between tiling spaces is a homeomorphism that is also a factor map.*

Topological conjugacies preserve the structure of tiling spaces as dynamical systems – targets for the action of the translation group. Topological conjugacies preserve dynamical invariants such as the set of translation-invariant probability measures, the dynamical spectrum, and mixing properties.

The strongest notion of equivalence is Mutual Local Derivability, or MLD. [**BSJ**]

DEFINITION. *Two tiling spaces are MLD if there is a topological conjugacy between them that is defined* locally. *More precisely, there exists a radius R such that, whenever two tilings T_1, T_2 agree on a ball of radius R around x, then $f(T_1)$ and $f(T_2)$ agree on a ball of radius 1 around x. Since f commutes with translations, it is sufficient to check this at $x = 0$.*

That is, the properties of $f(T)$ near the point x are determined by the properties of T on some ball around x. The term MLD was originally defined for tilings, rather than for tiling spaces.

DEFINITION. *If T and T' are tilings, we say that T' is* locally derivable *from T if, for some finite radius R, the properties of T' at each point x are determined by the properties of T in a ball of radius R around x. Stated formally, if there exists a radius R such that, whenever points x and y have the property that $[B_R + x]_T = [B_R + y]_T + y - x$, then $[B_1 + x]_{T'} = [B_1 + y]_{T'} + y - x$. If T' is locally derivable from T and T is locally derivable from T', then T and T' are MLD.*

Each Penrose chicken tiling is MLD to a Penrose kite-and-dart tiling. The periodic "all white" tiling of $\mathbb{R}$ is locally derivable from "one black tile", but "one black tile" is not locally derivable from "all white".

If the tilings T and T' are MLD, then the tiling spaces Ω_T and Ω'_T are automatically MLD. Just let $f(T) = T'$ and extend this to the orbit of T by $f(T + x) = T' + x$. The map f is uniformly continuous on $\mathcal{O}(T)$. Given ϵ, pick $\delta < (\epsilon^{-1} + R)^{-1}$. If $T - x$ and $T - y$ are δ-close, then the patterns in T around x and y agree (up to translation) out to distance δ^{-1}, so the patterns of T' around x and y agree (up to the same translations) up to distance $\delta^{-1} - R > \epsilon^{-1}$, so $f(T - x) = T' - x$ and $f(T - y) = T' - y$ are ϵ-close. We can therefore extend the map f to all of Ω_T by continuity. Conversely, if Ω and Ω' are tiling spaces that are MLD with map $f : \Omega \to \Omega'$, and if $T \in \Omega$, then the tilings T and $f(T)$ are MLD.

For a long time, it was believed that topologically conjugate tiling spaces were automatically MLD. In 1999, however, two independent papers [**Pet, RS**] showed that this was false. In Chapter 3 we will see how changing the shapes and sizes of tiles can sometimes yield spaces that are topologically conjugate but not MLD.

1.4. Contructing interesting tilings

There are three classes of tilings that come up repeatedly in tiling theory. First, there are *substitution tilings*. These are closely related to *self-similar* tilings, *pseudo-self-similar* tilings, and (pseudo) *self-affine* tilings. One dimensional examples include the Thue-Morse and Fibonacci tilings (see Figure 1.6), and the period-doubling tiling. Two dimensional examples include the chair tiling, the "table" or "domino" tiling, the half-hex, the Penrose tiling, and the pinwheel tiling. Most of the examples in this book will be substitution tilings.

Our second class of tilings also goes by many names. *Cut-and-project* tilings are closely related to *model sets*, to *canonical projection tilings*, and to *diffractive sets*. Examples include the Fibonacci tiling in one dimension, the Penrose and octagonal tilings in two dimensions, and the icosahedral tiling in three dimensions. This class of tilings overlaps with the substitution tilings, but neither is a subset of the other. There are plenty of substitution tilings that are not cut-and-project, and plenty of cut-and-project tilings that do not come from a substitution.

The third class of tilings are those defined by local matching rules. Imagine being given a box of tiles and having to tile the plane in any manner that fits. The tilings that result form a space that may or may not be the hull of a single tiling.

Substitutions in one dimension. To understand substitution tilings, we begin in one dimension. Pick a finite set (or "alphabet") $\mathcal{A}$, whose elements are called "letters". For instance, we might take $\mathcal{A} = \{a, b\}$. A sequence of letters is called a "word". We associate a word to each letter. (E.g., in the Thue-Morse tiling we associate $a \mapsto ab$, $b \mapsto ba$.) The substitution σ maps words to words, replacing each letter with its associated word. For instance, the Thue-Morse substitution has $\sigma(abbab) = abbabaabba$.

DEFINITION. *The substitution matrix M keeps track of the populations of different letters, with M_{ij} equaling the number of times that the i-th letter appears in $\sigma(j$-th letter).*

For example, the Thue-Morse substitution has $M = \left(\begin{smallmatrix} 1 & 1 \\ 1 & 1 \end{smallmatrix}\right)$. In general, $(M^n)_{ij}$ is the number of times that the i-th letter appears in $\sigma^n(j$-th letter).

DEFINITION. *A (substitution) matrix M is* primitive *if there is a positive integer n such that every entry of M^n is positive. Equivalently, there is an n such that $\sigma^n(any\ letter)$ contains every letter at least once. In such cases we will say that σ is a* primitive *substitution.*

THEOREM 1.2 (Perron-Frobenius). *Let M be a primitive matrix, and let λ_{PF} be its largest positive real eigenvalue (sometimes called the Perron-Frobenius eigenvalue). This eigenvalue has multiplicity one, and the corresponding right- and left-eigenvectors have strictly positive entries. All other eigenvalues have norm strictly less than λ_{PF}.*

For Thue-Morse, $\lambda_{PF} = 2$, the left-eigenvector is $L = (L_1, L_2) = (1, 1)$, and the right-eigenvector is $R = \left(\begin{smallmatrix} R_1 \\ R_2 \end{smallmatrix}\right) = \left(\begin{smallmatrix} 1 \\ 1 \end{smallmatrix}\right)$.

DEFINITION. *If σ is a primitive substitution on an alphabet $\{a_1, \ldots, a_k\}$, we call a bi-infinite word σ-admissible if each finite sub-word can be found in $\sigma^n(a_1)$ for some $n \geq 0$.*

In such cases, let $L = (L_1, \ldots, L_k)$ be a left-eigenvector of the substitution matrix corresponding to λ_{PF}, and consider tiles with labels $a_1, \ldots, a_k$ and lengths $L_1, \ldots, L_k$.

EXERCISE 1.3. *Show that the length of the patch corresponding to $\sigma^n(a_j)$ is $\lambda_{PF}^n L_j$.*

EXERCISE 1.4. *Show that, as $n \to \infty$, the fraction of letters of type a_i in $\sigma^n(a_j)$ approaches $R_i / \sum_k R_k$, and in particular is independent of j.*

DEFINITION. *Given a substitution σ, the sequence space S_σ is the set of all bi-infinite σ-admissible words and the tiling space Ω_σ consists of all tilings by the tiles $(a_1, \ldots, a_k)$ such that the corresponding sequence of letters $(a_1, \ldots, a_k)$ forms a σ-admissible word. The substitution σ acts on S_σ by replacing each letter with a word. It acts on Ω_σ by first stretching the tiling by a factor of λ_{PF} about the origin, and then replacing each stretched tile of type a_j with tiles corresponding to the word $\sigma(a_j)$.*

EXERCISE 1.5. *Show that for any integer $k > 0$ and any substitution σ, $\Omega_\sigma = \Omega_{\sigma^k}$*

DEFINITION. *A patch of a tiling corresponding to $\sigma^n(a_i)$ is called a supertile of level n (or order n) and type i. Ω_σ is the set of tilings with the property that every patch is found inside a supertile of some order.*

THEOREM 1.3. [**Mos, Sol**] *If σ is a primitive substitution and Ω_σ contains at least one non-periodic tiling, then every tiling in Ω_σ is non-periodic and $\sigma : \Omega_\sigma \to \Omega_\sigma$ is a homeomorphism.*

In particular, if $T \in \Omega_\sigma$, then there is a unique way to group the tiles of T into supertiles of order 1, such that the pattern of supertiles looks like a scaled-up version of a tiling in Ω_σ.

Examples.

(1) Thue-Morse sequences have an alphabet of two letters, a and b, with the substitution $\sigma(a) = ab$, $\sigma(b) = ba$. Note that $\sigma^2(a) = abba$ begins with a and $\sigma^2(b) = baab$ ends with b. Let $w = b.a$, with the dot indicating the location of the origin. $\sigma^2(w) = baab.abba$ contains w in the center. Likewise, $\sigma^4(w)$ contains $\sigma^2(w)$, and generally $\sigma^{n+2}(w)$ contains $\sigma^n(w)$. Taking limits we find a sequence $u = \ldots abbaabbabaab.abbabaabbaababba \ldots$ with $\sigma^2(u) = u$. We can make a tiling T out of u by thinking of each letter as a tile of length 1.

   EXERCISE 1.6. *Consider the patch bbaababbabaababbaabbaba of a Thue-Morse tiling. Group the tiles into supertiles of level 1, 2, etc., as far as you can go. Note that near the edges, a supertile may only be partially in the patch.*

EXERCISE 1.7. *Show that for the Thue-Morse substitution, $\Omega_\sigma = \Omega_T$.*

(2) The Fibonacci substitution is $\sigma(a) = b$, $\sigma(b) = ab$. The substitution matrix is $\left(\begin{smallmatrix} 0 & 1 \\ 1 & 1 \end{smallmatrix}\right)$, whose Perron-Frobenius eigenvalue equals the golden mean $\tau = (1 + \sqrt{5})/2$. The corresponding left- and right-eigenvectors are $L = (1, \tau)$ and $R = \left(\begin{smallmatrix} 1 \\ \tau \end{smallmatrix}\right)$. As with the Thue-Morse sequence, there is a fixed point of σ^2 built from the seed $b.a$. To get a tiling, we take the a-tile to have length 1 and the b-tile to have length τ. On average, there are τ b-tiles for every a-tile. This shows that there are no periodic tilings in Ω_σ, since the ratio of a to b-tiles in a periodic tiling would have to be rational.

(3) The period-doubling substitution has $\sigma(a) = bb$, $\sigma(b) = ab$. The matrix is $\left(\begin{smallmatrix} 0 & 1 \\ 2 & 1 \end{smallmatrix}\right)$, with $\lambda_{PF} = 2$, $L = (1, 1)$ and $R = \left(\begin{smallmatrix} 1 \\ 2 \end{smallmatrix}\right)$. Once again, a fixed point of σ^2 can be built from the seed $b.a$

There is a map from the Thue-Morse tiling space to the period-doubling tiling space. If a Thue-Morse tile is preceded by a tile of the same type, replace it with a period-doubling a tile. If it is preceded by a tile of the opposite type, replace it with a period-doubling b tile.

EXERCISE 1.8. *Let Ω_{TM} and Ω_{PD} denote the Thue-Morse and period-doubling substitution tiling spaces, respectively, and let $f : \Omega_{TM} \to \Omega_{PD}$ be as above. Show that f intertwines the two substitutions. That is, $\sigma_{PD} \circ f = f \circ \sigma_{TM}$. Use this fact to show that f is a 2:1 cover of Ω_{PD} by Ω_{TM}.*

A brief digression into history. Substitution *sequences* were studied at length long before aperiodic tilings became fashionable. Thue invented the Thue-Morse sequence in the late 1800s, and Morse reinvented it in the 1930s to prove properties of geodesics on Riemann surfaces. Traditionally, the central object of study wasn't the space of sequences, but rather a particular sequence that was fixed by some power of the substitution. The sequence space could then be recovered from this fixed point by taking its orbit under a shift map, and then taking the completion of that orbit in a metric that is similar to our tiling metric.

Substitution tilings in higher dimensions. A substitution in one dimension is combinatorial — replace each letter with a word. In higher dimensions, we must also consider the geometry of the tiles. Given a stretching factor $\lambda > 1$, a substitution is an operation that

(1) Stretches each tile by a linear factor λ, and
(2) Replaces each stretched tile by a cluster of (ordinary-sized) tiles, as in Figure 1.10. As before, these clusters are called supertiles (of order 1).

The substitution matrix is as before, and M_{ij} gives the number of i-tiles in a substituted j-tile. The left-eigenvector $L = (L_1, \ldots, L_k)$ specifies their volumes of the different tile types, but has no information about their shapes.

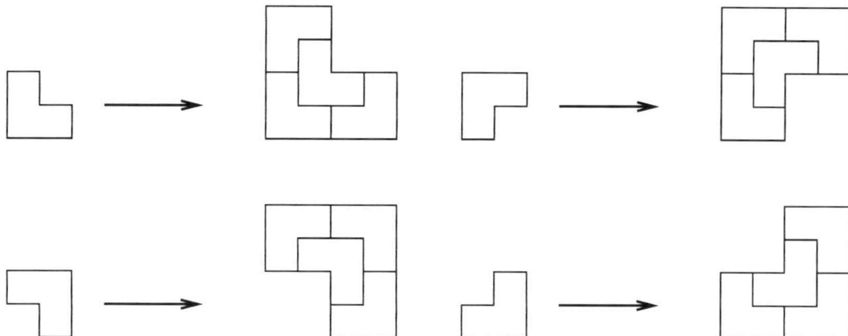

FIGURE 1.10. The chair substitution

We then define the tiling space

$$\Omega_\sigma = \{\text{Tilings } T | \text{ every patch of } T \text{ is found in a supertile of some order}\}$$
$$(1.1)$$

THEOREM 1.4. *If the substitution matrix is primitive, then there exists a fixed point of σ^n for some $n > 0$. (Such fixed points are called "self-similar tilings".) In particular, Ω_σ is non-empty.*

PROOF. If σ is primitive, there exists an integer $k > 0$ for which each supertile of order k and type a contains a copy of the tile a. By taking k large enough, we can assume that the supertile contains a copy of a *in its interior.* Let t_1 be the set of points in our supertile, and let t_2 be the chosen tile of type a inside t_1. We will find a point x inside t_2 in such a way that $(t_1 - x) = \lambda^k(t_2 - x)$. This means that $t_2 - x$ can be used as a seed for a self-similar tiling.

To find the point p, we subdivide the tile t_2 into smaller subtiles along the same pattern by which the supertile t_1 is subdivided into tiles. Let t_3 be the subtile that sits inside t_2 in the same way that t_2 sits inside t_1. Repeat the process indefinitely, subdividing t_n and letting t_{n+1} sit inside t_n the way that t_n sits inside t_{n-1}. The intersection of the nested sequence of subtiles t_n is nonempty and has diameter 0, hence is a single point, which we call x. $\square$

This procedure is illustrated for the chair tiling in Figure 1.11. If a is the chair in standard orientation (a square with the northeast corner missing), then $\sigma(a)$ contains two copies of a, but neither are in the interior. We have to go to $\sigma^2(a)$ to get interior tiles, and there are two copies of a in the interior of $\sigma^2(a)$, both of which are shaded. Using them as seed tiles, we can get two different self-similar tilings, one of which is shown in Figure 1.12.

Why did we need interior tiles? If we had chosen t_2 from $\sigma(a)$, then the point x would have been on the boundary of t_2, as in Figure 1.13. Using that as a seed would have yielded an infinite self-similar structure, but it would have covered only part of the plane. To get the whole plane we must

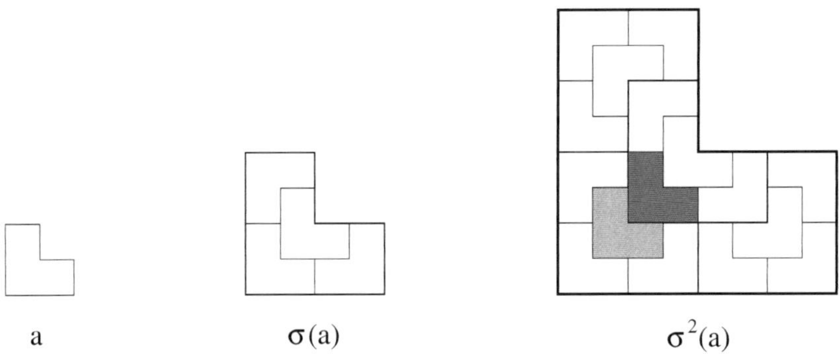

a   σ(a)   σ²(a)

FIGURE 1.11. There are two interior tiles of type a in $\sigma^2(a)$, but none in $\sigma(a)$.

FIGURE 1.12. A self-similar chair tiling

take x in the interior of t_2, which in turn means taking t_2 in the interior of t_1.

Historically, substitution tilings were often studied by looking at self-similar tilings and then considering the hulls of these tilings. The following theorem, combined with the existence of self-similar tilings, shows that this approach yields exactly the same spaces as we have been considering.

THEOREM 1.5. *If σ is a primitive substitution and $T \in \Omega_\sigma$, then $\mathcal{O}(T)$ is dense in Ω_σ, and $\Omega_\sigma = \Omega_T$.*

SKETCH OF PROOF. We must show that if $T' \in \Omega_\sigma$, then every patch of T' is found somewhere in T. To do this, we show that for each tile t and integer $n \geq 0$, there exists a radius R such that every ball of radius R in T contains a copy of $\sigma^n(t)$. For $n = 0$ this comes from the primitivity of σ. For large values of n it follows from the $n = 0$ result combined with the fact that σ is invertible on Ω_σ. $\qquad\square$

EXERCISE 1.9. *Expand this sketch into a complete proof.*

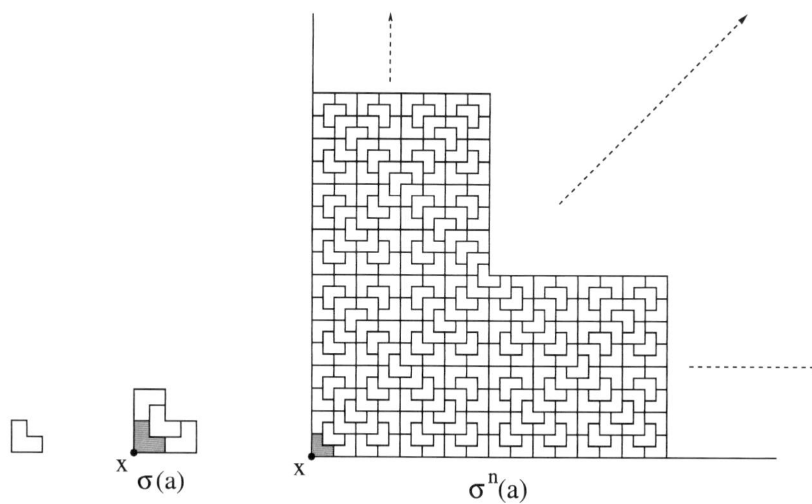

FIGURE 1.13. Starting from a corner seed, we only tile a quarter plane.

There are several natural extensions of the notions of self-similarity and of substitution tilings. If T is a tiling and T is locally derivable from a stretched version of T, then T is called "pseudo-self-similar". This is somewhat weaker than self-similarity, since the tiles of T may not group exactly into the tiles of λT. However, Priebe and Solomyak [**PS**] have proven that every pseudo-self-similar tiling of the plane is MLD to a self-similar tiling, albeit one whose tiles may have fractal boundaries. In dimensions greater than one, we sometimes consider expansive linear maps instead of just rescaling. If L is an expansive linear transformation and $L(T)$ can be subdivided into T, then we say L is "self-affine", and if T is locally derivable from $L(T)$ we say that T is "pseudo-self-affine". In the literature, the term "substitution tiling" is sometimes used for each of these generalizations, and "substitution tiling space" is sometimes used for the hulls of such tilings. In the bulk of this book, however, substitutions will be limited to rescalings by a factor $\lambda > 1$ followed by subdivision into geometric pieces.

Several famous planar substitutions are shown in Figure 1.14.

Cut-and-project tilings. A k-dimensional cut-and-project tiling is obtained by taking a periodic structure in $\mathbb{R}^n$, restricting it to the neighborhood of a k-plane in $\mathbb{R}^n$ (that's the "cut"), and then projecting the periodic structure onto that k-plane. For generic embeddings of the k-plane in $\mathbb{R}^n$, the result will be an aperiodic tiling of the k-plane. By considering all translates (in $\mathbb{R}^n$) of the periodic structure, we can construct a space of cut-and-project tilings. The topological and dynamical properties of this space are closely related to those of the n-torus and to the method of cutting.

We illustrate the method with a family of one-dimensional tilings obtained by projecting from $\mathbb{R}^2$. Pick an irrational number $\alpha > 0$ and a point

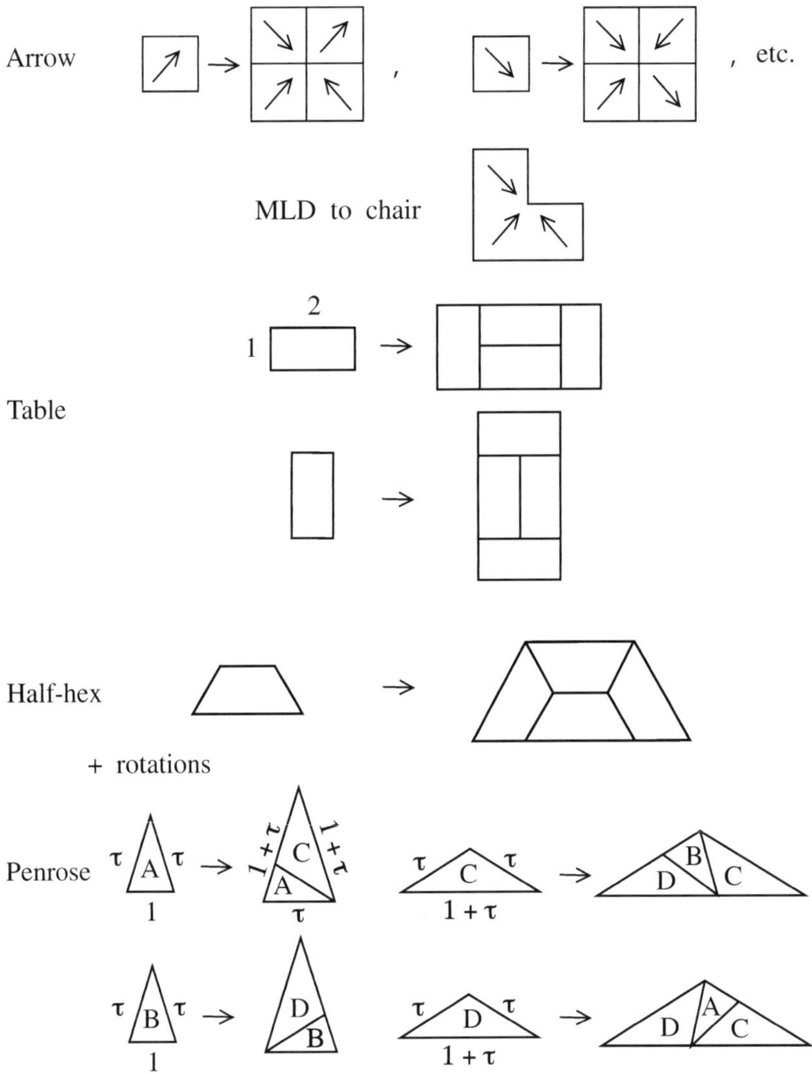

FIGURE 1.14. The arrow, table, half-hex, and Penrose substitutions. The letter τ denotes the golden mean $(1 + \sqrt{5})/2$.

$p = (x_0, y_0) \in R^2$, as in Figure 1.15. Let L be the line through p with slope α. Let S be the strip obtained by adding points in the unit square to points in L. Equivalently, it is the union of all the unit squares with southwest corner on L.

Now let $\Lambda = \mathbb{Z}^2$ be the set of points in $\mathbb{R}^2$ with integer coordinates, let $\Lambda_1 = \Lambda \cap S$, and let Λ_2 be the orthogonal projection of Λ_1 onto L. This gives a point pattern on L. We then take our tiles to be the intervals between successive points. These either have length $1/\sqrt{1 + \alpha^2}$ if the successive points were the projections of (m, n) and $(m + 1, n)$, or length $\alpha/\sqrt{1 + \alpha^2}$ if the

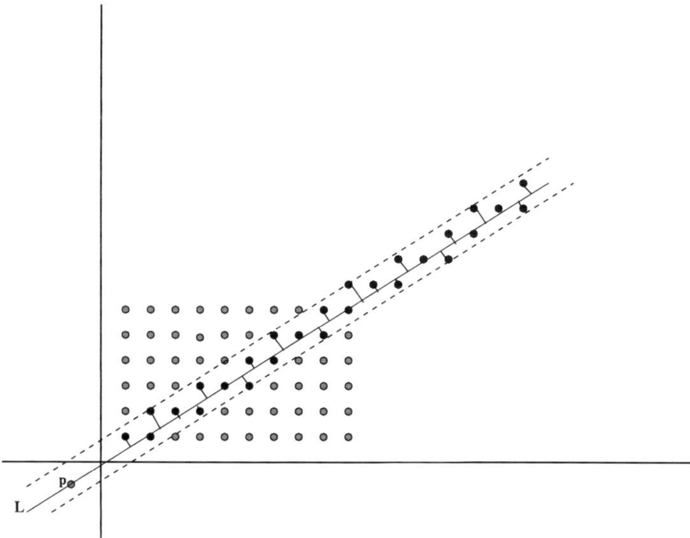

FIGURE 1.15. Projecting from two dimensions to one

successive points were projections of (m, n) and $(m + 1, n)$. We can then identify L with the real line, with p corresponding to the origin.

The perceptive reader will have noticed that I did not say whether S was constructed from the open unit square or the closed unit square. If L goes through an integer point (m, n), then both $(m + 1, n)$ and $(m, n + 1)$ will be on the boundary of S. Do we include one point, the other, or both in Λ_1? The set of points p for which this problem arises has measure zero, but is still infinite. For each irrational α, let Ω_α^0 be the set of tilings constructed this way in which L does not contain any integer points. Then let Ω_α be the completion of Ω_α^0 in the tiling metric.

If m and n are integers, then the tiling obtained from $p = (x_0 + m, y_0 + n)$ is exactly the same as that obtained from $p = (x_0, y_0)$. Conversely, if (x_0, y_0) and (x_1, y_1) do not differ by an integer vector, then the tilings built from these two choices of p are not the same. This implies that Ω_α^0 is isomorphic to the 2-torus, minus the points corresponding to lines L that hit integers. This set of singular lines is itself a line in the 2-torus, albeit one that winds densely.

To understand the points of Ω_α that are not in Ω_α^0, we consider tilings based on the point $p = (m + \epsilon, n)$ in the limit of $\epsilon \to 0$. As $\epsilon \to 0^+$, the tilings converge in the tiling metric and contain the projection of the point $(m + 1, n)$, but not the projection of $(m, n + 1)$. As $\epsilon \to 0^-$, the tilings contain the projection of $(m, n + 1)$, but not that of $(m + 1, n)$. The two limits are otherwise identical. The upshot is that there are *two* tilings in Ω_α that correspond to $p = (m, n)$, and these points are not close in the tiling metric! Topologically, the space Ω_α consists of the 2-torus with an irrational line removed and then glued in twice — once as the limit of one side and once as the limit of the other.

The space Ω_α can be constructed for any irrational α. It is true (but not obvious!) that when α is a quadratic irrational number, then Ω_α can also be constructed from a substitution. In particular, the space of tilings with $\alpha = \tau = (1 + \sqrt{5})/2$ is none other than the Fibonacci tiling space.

There are many ways to generalize this construction. The strip does not have to come from a unit square, but can take the form $L \times E$, where the "window" E is any (reasonable) set in the orthogonal complement to L. The source space does not have to be $\mathbb{R}^n$, but instead can be any locally compact Abelian group. For instance, the Penrose tiling is most naturally described as a projection from $\mathbb{R}^4 \times \mathbb{Z}_5$ to $\mathbb{R}^2$. For a more detailed description of cut-and-project tilings, their generalizations, and the topology of the tiling spaces they generate, see [**Moo**].

Topologically, a space of cut-and-project tilings is always a torus in the higher-dimensional space, minus a lower-dimensional piece where an integer point lies on the boundary of the strip S, plus limits as those singular points are approached from various directions. Forest, Hunton and Kellendonk have developed powerful methods for computing topological invariants of such a space in terms of the geometry of the window E, and have written an excellent book on the subject [**FHK**]. These techniques have a very different flavor from the other content of this book, and will not be discussed further.

Local matching rules. The constructions of substitution tilings and cut-and-project tilings are highly non-local. For substitution tilings, we apply a condition to patches of arbitrarily large size. For cut-and-project tilings, we invoke a higher dimension and project globally. At first glance, neither would seem to model actual materials, where atoms and molecules are held together by local forces.

A different approach is to take a set S of tiles together with rules about how two tiles can fit together. Usually these rules can be implemented geometrically, by adding bumps and notches to the tiles, as in Figure 1.16. Let Ω_S be the set of all tilings of the plane (or $\mathbb{R}^d$) by translates of the tiles in S. If Ω_S is non-empty and if every tiling in Ω_S is aperiodic, we say that S is an *aperiodic set*.

Tiling theory largely arose from a decidability question: given a set of tiles, is it possible to determine algorithmically whether or not they tile the plane? It was generally believed that any set that tiled the plane could tile the plane periodically, and it was possible to determine whether a set of tiles could produce a periodic tiling. However, it turned out that some sets of tiles do tile the plane, but only non-periodically! The first aperiodic set, by Berger [**Ber**], had thousands of tiles. Simpler sets were later constructed by Robinson [**Rob**], by Kari and Culik [**Kar, Cul**], and by others. The most famous aperiodic set contains the Penrose kites and darts of Figure 1.16: two shapes in ten orientations, or twenty tiles in all.

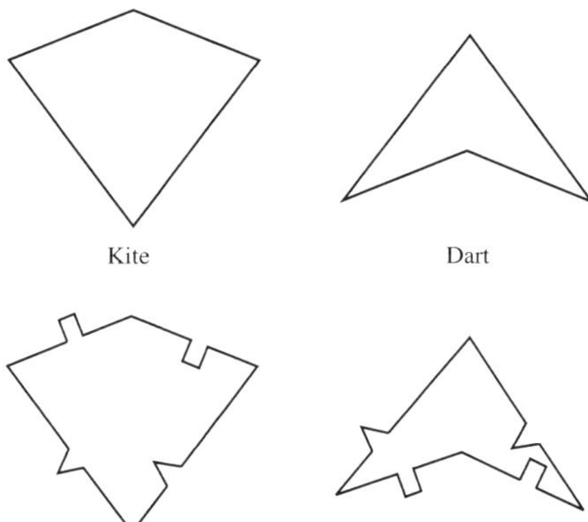

FIGURE 1.16. Penrose kite and dart tiles, with bumps to enforce matching rules

A remarkable theorem of Shahar Mozes [**Moz**] says that for any substitution σ involving square tiles of equal size, there exists a set S and a factor map $f : \Omega_S \to \Omega_\sigma$ that is onto, measure-preserving and 1:1 away from a negligible set (meaning a set of measure zero with respect to *all* translation-invariant probability measures). Mozes' argument was generalized by Radin [**Rad**] to apply to the pinwheel tiling, and Mozes' theorem was generalized by Goodman-Strauss [**GS**] to apply to any substitution tiling (as defined in this book).

Despite their historical significance, and despite Goodman-Strauss' theorem, we will not be talking much about matching rules tiling spaces. While powerful techniques have been developed to explore the topology of substitution and cut-and-project tiling spaces, at this time the topology of most local matching rules tiling spaces remains a mystery.

CHAPTER 2

Tiling spaces and inverse limits

In Chapter 1 we defined a variety of tilings and tiling spaces using substitutions, cut-and-project methods, and local matching rules. In this chapter we begin to study the the topology of such spaces, and relate tiling spaces to objects that are better-known, namely inverse limits of branched manifolds.

2.1. Local structure

Suppose Ω is a tiling space and $T \in \Omega$ is a tiling. What does an ϵ-neighborhood of T look like? If T' is in that ϵ-neighborhood, then T and T' agree on $B_{1/\epsilon}$, up to a small translation. In other words, all the tilings in the neighborhood can be obtained by the following procedure:

(1) Take the patch $[B_{1/\epsilon}]_T$.
(2) Wiggle this patch by up to ϵ in any direction. This gives our tiling space continuous degrees of freedom.
(3) Extend the patch out to infinity. This involves discrete choices.

If, as is typical, there are infinitely many discrete choices to be made, the set of such choices will be uncountable – typically a Cantor set. In other words, Ω looks locally like the product of a d-dimensional disk (if we are tiling $\mathbb{R}^d$) and a Cantor set. Such spaces are old hat to people who work in ergodic theory and dynamical systems, but are pretty weird to those of us who learned topology in the context of smooth manifolds! Fortunately, inverse limit spaces provide plenty of examples with which to stretch our intuition.

2.2. Inverse limit spaces

Suppose $\Gamma_0, \Gamma_1, \ldots$ are topological spaces, and for each non-negative integer n, $f_n : \Gamma_{n+1} \to \Gamma_n$ is a continuous map. Consider the product space $\Pi \Gamma_i$, with the product topology. This can be viewed as a set of sequences $(x_0, x_1, \ldots)$ with each $x_n \in \Gamma_n$. We define the *inverse limit space*

$$\varprojlim(\Gamma, f) = \{(x_0, x_1, \ldots) \in \Pi\,\Gamma_n | \text{for all } n,\ x_n = f_n(x_{n+1})\}. \qquad (2.1)$$

The spaces Γ_n are called *approximants* to the inverse limit since, if you know x_n, you automatically know $x_0, \ldots, x_{n-1}$. We are working in the product topology, where two sequences $(x_0, \ldots)$ and $(y_0, \ldots)$ are close if their first N terms are close for a large value of N. If the sequences are close, then x_N and y_N are close. Conversely, if x_N and y_N are sufficiently

21

close, then by the continuity of f_{N-1}, x_{N-1} and y_{N-1} are close, and likewise each x_i with $i < N$ is close to y_i.

A simple example of an inverse limit space is the dyadic solenoid. Let each Γ_n be the unit circle $\mathbb{R}/\mathbb{Z}$, and let each f_n be multiplication by 2. That is, Γ_1 is a circle that wraps twice around Γ_0, Γ_2 is a circle that wraps twice around Γ_1, and so on, as in Figure 2.1. A point in $\varprojlim(\Gamma, f)$ is described by a point x_0 on the circle Γ_0 (a continuous degree of freedom), a choice of x_1 from among the two preimages of x_0, a choice of x_2 from among the two preimages of x_1, etc. Our space is locally the product of an interval and a countable number of 2-fold choices, i.e., the product of an interval and a Cantor set.

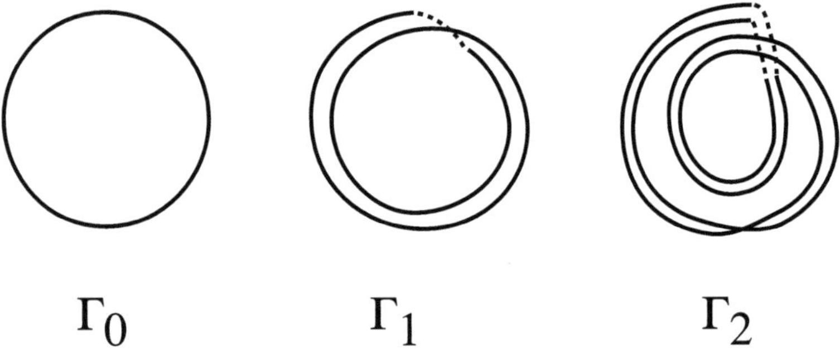

FIGURE 2.1. The first three approximants to the dyadic solenoid

2.3. Tiling spaces are inverse limits

In the past decade, there has been an explosion of work of the topology of tilings, most of which relies on constructions of tiling spaces as inverse limits of simpler spaces. This explosion was begun in 1998 by Anderson and Putnam [**AP**], who showed that substitution tiling spaces are inverse limits of branched manifolds. We will review the Anderson-Putnam construction later in this chapter.

Ormes, Radin and Sadun [**ORS**] generalized the Anderson-Putnam construction to allow for pinwheel-like tilings that have tiles pointing in arbitrary directions. Bellissard, Benedetti and Gambaudo [**BBG**] gave a very general construction that applied to the hull of any (simple) tiling, not just a substitution tiling. They then used this construction to prove powerful gap-labeling theorems. Independently, Franz Gähler [**Gah**] gave an exceptionally simple proof that hulls of simple tilings are inverse limit spaces. Benedetti and Gambaudo [**BG**] then generalized the Bellissard, Benedetti and Gambaudo construction, and Sadun [**Sa2**] independently generalized the Gähler construction, both efforts including pinwheel-like tilings and tilings of non-Euclidean spaces. Recent work by Frank and Sadun [**FS**] shows that even spaces without "finite local complexity" (see chapter 7) can be understood using inverse limits.

2.4. Gähler's construction

Let Γ_0 be the set of all possible instructions for laying a tile at the origin, let Γ_1 be instructions for laying that center tile plus a ring of tiles around it, and generally let Γ_n be instructions for laying down a center tile and n layers of tiles around it. The map $f_n : \Gamma_{n+1} \to \Gamma_n$ is just the "forgetful map" that forgets the instruction for the outermost ring. The statement that $x_n = f_n(x_{n+1})$ just says that the instructions x_n and x_{n+1} agree on the first n rings.

A point in the inverse limit is a nested sequence of consistent instructions for placing a tile at the origin and an infinite number of rings around it. In other words, it is instructions for laying out an entire tiling. The inverse limit, being the set of all such instructions, is naturally isomorphic to the tiling space itself.

The only problem with this argument is that sets of instructions aren't topological spaces! We need to show that Γ_n can be viewed as a geometric object whose topological properties are computable. We begin with Γ_0.

A point in Γ_0 tells us how to place a tile at the origin. This involves a choice of tile, followed by a choice of a point in that tile to align with the origin. In other words, Γ_0 contains one copy, and only one copy, of every kind of tile that is allowed.

But what happens if the origin lies on an edge? In giving instructions for the patch shown in Figure 2.2a, do we specify a point on the right edge of tile A or a point on the left edge of tile B? The answer is to identify the right edge of A with the left edge of B! That is, Γ_0 is the union of one copy of each tile type, with some edges identified. If somewhere in some tiling, a tile of type A shares an edge with a tile of type B, then those two edges are identified.

The result is a branched surface. If more than one kind of tile can appear to the left of B (say, A and C), then the left edge of B will be identified with both the right edge of A and with the right edge of C, as in Figure 2.2b.

Collaring. Suppose we have a tiling T. We can label the tiles of T not only by their own type, but by the pattern of their nearest neighbors. Tiles with such labels are called *collared tiles*. In Figure 2.3, tiles A and B are the same as uncollared tiles, but are considered different as collared tiles.

Strictly speaking, relabeling the tiles gives us a different set of possible tiles, and hence a different tiling. Call this tiling T', and let $\Gamma_1(T) = \Gamma_0(T')$. It is a branched manifold, as before. A point in $\Gamma_1(T)$ describes how to place a collared tile at the origin, which is equivalent to describing how to place a tile at the origin together with all of its nearest neighbors. Likewise, for any $n > 0$, we let $\Gamma_n(T) = \Gamma_{n-1}(T')$.

As an example, we construct Γ_0 and Γ_1 for the Fibonacci tiling. There are two (uncollared) tile types, a and b, so Γ_0 is the union of one a tile and one b tile, modulo identifications. Since a is always followed by b, the endpoint of the a tile is identified with the beginning of the b tile. Since b is

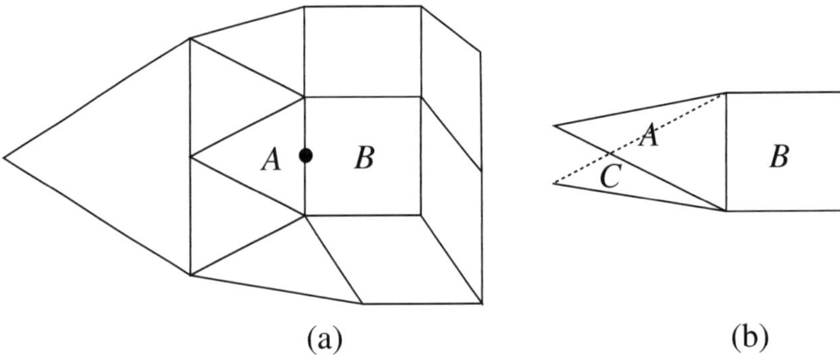

(a)   (b)

FIGURE 2.2. (a) What instructions do you give when the origin sits on an edge? (b) Gluing tiles yields a branched manifold.

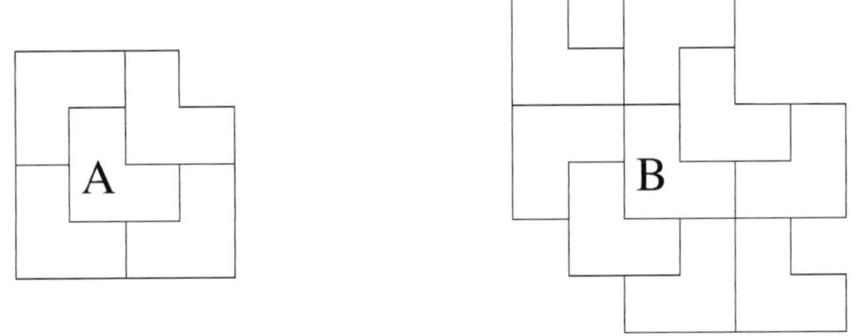

FIGURE 2.3. Collared tiles

sometimes followed by a, the endpoint of b is identified with the beginning of a. Finally, since b is sometimes followed by b, the beginning and end of b are identified. The result is a figure-eight, as shown in Figure 2.4.

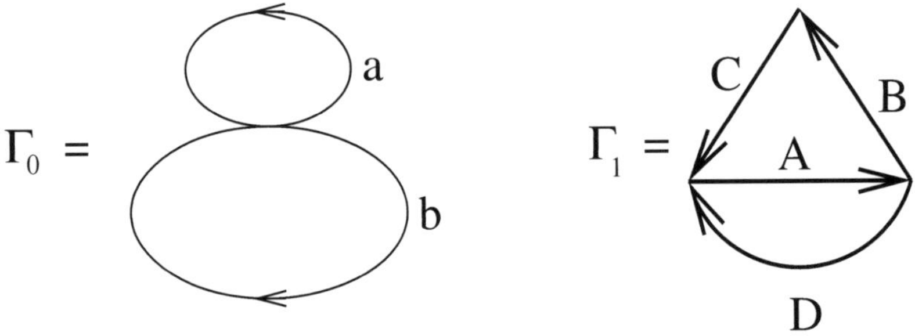

FIGURE 2.4. Two Gähler approximants for the Fibonacci tiling space

Suppose we have a Fibonacci tiling

$$T = \ldots abbabbababbab \ldots \tag{2.2}$$

Notice that each a tile is preceded and followed by a b tile. We therefore define the collared tile $A = (b)a(b)$, where the parentheses indicate the tiles before or after the one in question. There are three collared b tiles, namely $B = (a)b(b)$, $C = (b)b(a)$, and $D = (a)b(a)$. (The patch bbb never appears in a Fibonacci tiling.) We can rewrite T as

$$T' = \ldots ABCABCADABCA \ldots \tag{2.3}$$

(The visible patch of T' is one tile shorter than that of T, since we do not know how to label the last visible b in T as a collared tile.) Since A is followed by B or D, B is always followed by C, and C and D are always followed by A, we obtain the complex Γ_1, as shown in Figure 2.4.

Note that this construction is not specific to the tiling T. Another tiling $\tilde{T}$ in the same space would have the same patches and the same set of collared tiles as T, and would yield the same complexes Γ_0 and Γ_1. The approximants Γ_n describe the entire tiling *space*.

2.5. The Anderson-Putnam construction

Gähler's construction is simple, elegant, and conceptually powerful. It has been used to prove a variety of important theorems about tiling spaces. When it comes to computing topological invariants, however, it is almost useless. The approximants Γ_n and the bonding maps f_n depend on n, and can get successively more complicated as n increases.

With substitution tilings it is possible to avoid this problem. Anderson and Putnam defined an inverse limit structure in which all of the Γ_n's were essentially the same and all of the f_n's were essentially the same. We will first lay out the Anderson-Putnam (AP) construction in the easiest cases, where the substitution has a property called "forcing the border". We will then show how to modify the construction to handle all substitutions.

Suppose we have a substitution σ with a linear stretching factor of λ. Let Γ_0 be exactly as before, with one copy of each tile type, stitched together at edges. Let Γ_1 be a larger version of Γ_0, namely Γ_0 stretched by a factor of λ. Similarly, let Γ_n be Γ_0 expanded by a factor of λ^n. We view Γ_n as containing one copy of each supertile of level n.

The approximant Γ_0 tells us how to place a tile at the origin. The approximant Γ_n tells us how to place a supertile of level n that contains the origin. There is a forgetful map $f_n : \Gamma_{n+1} \to \Gamma_n$ that restricts attention to the level-n supertile containing the origin, while forgetting about the rest of the level-$(n + 1)$ supertile containing the origin.

Figure 2.5 shows Γ_0, Γ_1 and Γ_2 for the Fibonacci tiling. The map f_0 sends the upper loop of Γ_1 to the lower loop of Γ_0, and sends the lower loop of Γ_1 to the upper loop of Γ_0 followed by the lower loop. Likewise, f_1 sends the upper loop of Γ_2 to the lower loop of Γ_1, and the lower loop of Γ_2 to the upper and lower loops of Γ_1. If we rescale the approximants Γ_n to make them all the same as Γ_0, then the maps f_n are all the same. Each f_n

is just the substitution map $\sigma : \Gamma_0 \to \Gamma_0$ that takes each tile, stretches it, subdivides it, and sends each piece to Γ_0.

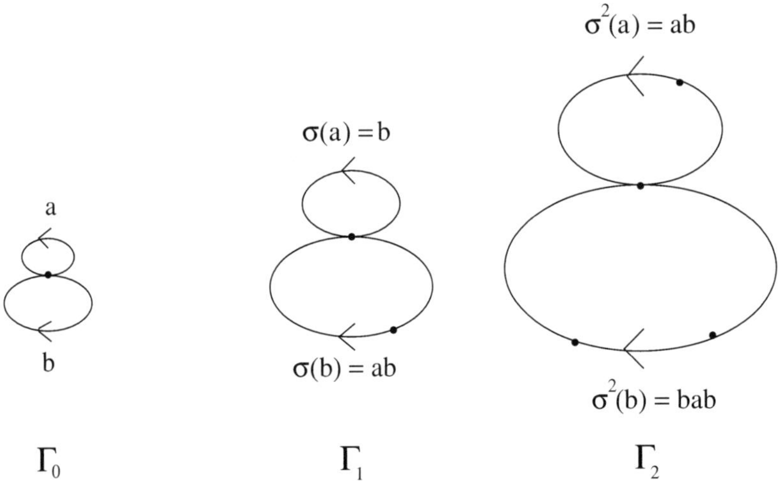

FIGURE 2.5. The first three Anderson-Putnam approximants to the Fibonacci tiling space

Topologically, $\varprojlim(\Gamma, f)$ is the inverse limit of a single branched manifold Γ_0 by a single map σ. A point in the inverse limit gives consistent instruction for placing the origin in a tile, that tile in a level-1 supertile, that supertile in a level-2 supertile, and so on forever. In other words, it tells how to put the origin in an infinite supertile. However, that infinite supertile may not cover the entire plane, as is seen in Figure 2.6! The infinite supertile can always be extended to a tiling of the entire plane, but that extension may not be unique.

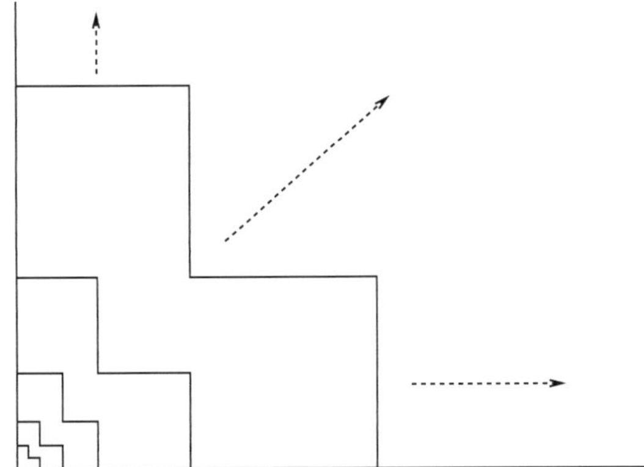

FIGURE 2.6. An infinite-order supertile may not cover the entire plane.

Forcing the border. A substitution is said to *force the border* if there exists a positive integer n such that any two level-n supertiles of the same type have the same pattern of neighboring tiles. In that case, a level-$(n+k)$ supertile determines all of the level-k supertiles around it, and an infinite supertile determines the entire plane. Points in the inverse limit $\varprojlim(\Gamma, f)$ are then in 1-1 correspondence with tilings of the plane, and we have a topological isomorphism $\varprojlim(\Gamma, f) \leftrightarrow \Omega_\sigma$.

This phenomenon is best illustrated with the example of the half-hex tiling. In the half-hex tiling, each tile pairs off with another tile to form a hexagon. Knowing the position and type of one tile actually tells you about two: that tile and its partner. Knowing the position and type of a level-1 supertile tells you about the partner of that supertile, and also about the partners of the unpaired tiles within the two supertiles. Figure 2.7 shows, in dotted lines, the tiles determined by a half-hex tile, a level-1 supertile and a level-2 supertile. The level-2 supertile determines *all* of its immediate neighbors, so the half-hex substitution forces the border with $n = 2$.

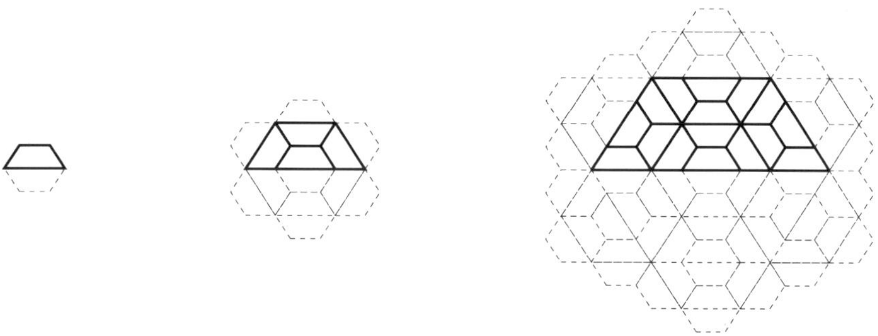

FIGURE 2.7. In bold face, a half-hex tile, an order-1 supertile, and an order-2 supertile. In dotted lines, the nearby tiles that these determine.

Among 1-dimensional substitutions, Thue-Morse does not force the border. The Fibonacci and period-doubling substitutions force the border on one side only; since all supertiles end in b, there is a b to the left of every supertile, but there is no way to determine the tile to the right of a general supertile. By contrast, the substitution $\sigma(a) = abaab$, $\sigma(b) = aabab$ does force the border, since all supertiles begin with a and end with b.

Among 2-dimensional substitutions, the Penrose substitution forces the border. In fact, the term "force the border" was coined by Kellendonk [**Kel1**] in his work on the Penrose tilings. The chair substitution does not force the border. The tiles along the edges of a large supertile are determined, but the tiles at the corner are not, as illustrated in Figure 2.8. The arrow and table substitutions also fail to force the border.

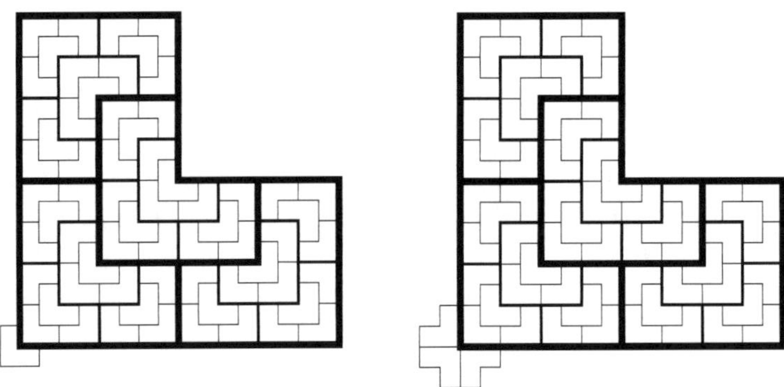

FIGURE 2.8. There are two ways to extend a high-order chair supertile around the southwest corner.

The Anderson-Putnam collaring trick. There is a simple trick, introduced by Anderson and Putnam [**AP**], for describing an arbitrary substitution tiling space by a substitution that forces the border. Just rewrite the substitution in terms of collared tiles!

We illustrate the method with the Fibonacci substitution $\sigma(a) = b$, $\sigma(b) = ab$ and the Fibonacci collared tiles $A = (b)a(b)$, $B = (a)b(b)$, $C = (b)b(a)$ and $D = (a)b(a)$. Substituting an a that is preceded and followed by a b gives a b that is preceded and followed by the word ab. In terms of collared tiles, that is an A preceded by a B and followed by an A. Applying the same reasoning to all four collared tiles we get

$$
\begin{aligned}
\sigma(A) &= (ab)b(ab) = (B)C(A) \\
\sigma(B) &= (b)ab(ab) = AD(A) \\
\sigma(C) &= (ab)ab(b) = (D)AB \\
\sigma(D) &= (b)ab(b) = AB \\
\sigma^2(A) &= (AD)AB(C) \\
\sigma^2(B) &= (B)CAB(C) \\
\sigma^2(C) &= (AB)CAD(A) \\
\sigma^2(D) &= (C)CAD(A)
\end{aligned}
\tag{2.4}
$$

This rewritten substitution does not force the border with $n = 1$. We know the *uncollared* tile that precedes $\sigma(B)$, namely b, but we do not know the *collared* tile that precedes $\sigma(B)$ (or those that precede $\sigma(D)$ or follow $\sigma(C)$ or $\sigma(D)$). However, it does force the border with $n = 2$.

EXERCISE 2.1. *Prove that a substitution applied to collared tiles always forces the border. Do this first in one dimension, and then in higher dimensions.*

We summarize the situation with the theorem proved by Anderson and Putnam:

THEOREM 2.1. *Let Γ_0 be the cell complex obtained from stitching one copy of each (uncollared) tile, and let $\tilde{\Gamma}_0$ be the similar cell complex obtained* *by stitching one copy of each collared tile. (In the Gähler construction this was called Γ_1.) In both cases we denote the bonding map by σ, as it is induced by the substitution. Ω_σ is always homeomorphic to $\varprojlim(\tilde{\Gamma}_0, \sigma)$. If σ forces the border, then Ω_σ is also homeomorphic to $\varprojlim(\Gamma_0, \sigma)$.*

In principle, we could take the inverse limit of a cell complex built from twice-collared tiles, or 17-times collared tiles. However, that's just computational overkill. In fact, collaring once is often overkill.

The Fibonacci substitution forces the border on the left, so we don't need to collar our Fibonacci tiles on the left, just on the right. Likewise, the chair tiling forces the border along the edges, but not at the corners. To construct the chair tiling space, we need only collar our tiles in a way that preserves information about those corners. By choosing a suitable "partial collaring" scheme, one can often simplify computations considerably. We will revisit the question of efficient collaring in Chapter 6.

EXERCISE 2.2. *Relabel the tiles in the Fibonacci tiling so they are collared on the right but not on the left. Construct the partially collared Anderson-Putnam complex (call it $\Gamma_{1/2}$). What does it look like as a topological space? Describe explicitly the map $\sigma : \Gamma_{1/2} \to \Gamma_{1/2}$.*

EXERCISE 2.3. *Compute the Anderson-Putnam complexes Γ_0 and $\tilde{\Gamma}_0$ for the Thue-Morse substitution and the maps σ on each one. Unlike for the Fibonacci substitution, the two spaces are not homotopy equivalent, and the inverse limit of Γ_0 is not homeomorphic to the inverse limit of $\tilde{\Gamma}_0$.*

EXERCISE 2.4. *Devise an efficient collaring scheme for the arrow substitution. The fully collared complex $\tilde{\Gamma}_0$ contains $13 \times 4 = 52$ cells (13 patterns in 4 orientations), but it is possible to make do with only $6 \times 4 = 24$.*

CHAPTER 3

Cohomology of tilings spaces

Of all the topological invariants of tiling spaces, by far the best studied is the Čech cohomology. In large part this is because we know how to compute it. Anderson and Putnam showed how to write a tiling space as an inverse limit, and the cohomology of a inverse limit can be deduced from the cohomologies of the approximants. Much of this chapter is devoted to this technology. Cut-and-project tiling spaces are homotopy equivalent to tori with certain lower-dimensional sets removed. Forest, Hunton and Kellendonk used this description to compute the cohomology of such spaces in terms of the geometry of the "window".

Let's face it, however. "Because we can" isn't much of a justification for doing something! Better reasons for computing tiling cohomology are that the cohomology gives interesting information about tiling spaces, and that other topological invariants, like homology and fundamental group, simply don't work.

Suppose T is an aperiodic substitution tiling. Ω_T then has one connected component, but uncountably many path-components. Since π_0 of a space is the set of path components, and since H_0 is freely generated by π_0, these invariants are useless when applied to Ω_T. Higher homotopy groups probe a single path component, namely the one containing the base point. However, each path component in a tiling space is an orbit under $\mathbb{R}^d$. The orbit of an aperiodic tiling is contractible, so $\pi_n(\Omega_T) = 0$ for $n > 0$. Likewise, $H_n(\Omega_T) = 0$.

Čech cohomology does better. In dimension zero, Čech cohomology measures *connected* components, not path components, so $\check{H}^0(\Omega_T) = \mathbb{Z}$. Čech cohomology also behaves well under inverse limits. The Čech cohomology of the inverse limit is the direct limit of the Čech cohomology of the approximants:

$$\check{H}^*(\varprojlim(\Gamma, f)) = \varinjlim \check{H}^*(\Gamma) = \varinjlim H^*(\Gamma), \qquad (3.1)$$

with the last equality following from the fact that Čech and singular cohomology are isomorphic for CW complexes.

That's a very pretty formula, but so far we have only defined one of the three essential terms! We know what inverse limits are, but what does the Čech cohomology $\check{H}^*$ mean, and what does the direct limit $\varinjlim$ mean?

31

3.1. Direct limits

Suppose we have the following setup: We are given a family of groups G_α indexed by a partially ordered set I. I is a *directed set*, meaning that for any pair α, β of elements of I, there is a third element γ with $\alpha < \gamma$ and $\beta < \gamma$. For any pair α, γ with $\alpha < \gamma$, we have a homomorphism $\rho_{\gamma\alpha} : G_\alpha \to G_\gamma$. If $\alpha < \beta < \gamma$, then $\rho_{\gamma\alpha} = \rho_{\gamma\beta} \circ \rho_{\beta\alpha}$.

DEFINITION. *The direct limit of the groups G_α, denoted $\varinjlim(G_\alpha, \rho_{\beta\alpha})$ or $\varinjlim G_\alpha$, is the disjoint union of all the G_α modulo the following equivalence. If $\alpha < \gamma$ and $x \in G_\alpha$, then $x \sim \rho_{\gamma\alpha}x \in G_\gamma$. If $x \in G_\alpha$ and $y \in G_\beta$, and if $\alpha < \gamma$ and $\beta < \gamma$, then x and y can be identified with elements $\tilde{x}, \tilde{y} \in G_\gamma$, and the product xy is identified with $\tilde{x}\tilde{y} \in G_\gamma$.*

EXERCISE 3.1. *Show that the multiplication law is well-defined, and that $\varinjlim G_\alpha$ is a group.*

The same definition applies to rings, and the definition of group operations applies both to addition and multiplication. Here are some examples of direct limits of Abelian groups.

(1) Let I be the natural numbers and let each $G_\alpha = \mathbb{Z}$. The element n in G_α will be denoted by the pair (n, α). For $\gamma > \alpha$, let $\rho_{\gamma\alpha}$ be multiplication by $2^{\gamma-\alpha}$. That is, $(n, \alpha) \sim (2n, \alpha + 1)$, and we are taking the direct limit of

$$\mathbb{Z} \xrightarrow{\times 2} \mathbb{Z} \xrightarrow{\times 2} \mathbb{Z} \xrightarrow{\times 2} \cdots \tag{3.2}$$

The direct limit is isomorphic to a subset of the rational numbers, namely those whose denominators are powers of 2, by $(n, \alpha) \leftrightarrow n/2^\alpha$. These numbers are called *dyadic rationals*, and the set of dyadic rationals is denoted $\mathbb{Z}[1/2]$, since (as a set) is equals the ring obtained by adjoining $1/2$ to the integers.

(2) Next let I be the natural numbers, let each $G_\alpha = \mathbb{Z}$, and let each map $\rho_{\gamma\alpha}$ be the zero map. Since all elements get identified to zero, the direct limit is the trivial group. This shows that we need to consider the maps $\rho_{\gamma\alpha}$ as well as the groups G_α when computing a direct limit.

(3) Again, let I be the natural numbers. Let each $G_\alpha = \mathbb{Z}^2$, and let each $\rho_{\alpha+1,\alpha}$ be given by the matrix $\left(\begin{smallmatrix} 0 & 1 \\ 1 & 1 \end{smallmatrix}\right)$. Each map ρ is an isomorphism, so all the groups G_α are identified, and the direct limit is $\mathbb{Z}^2$. This direct limit group has an interesting linear ordering. We call an element $\left(\begin{smallmatrix} n_1 \\ n_2 \end{smallmatrix}\right) \in G_\alpha$ positive if $n_1 + \tau n_2 > 0$, where τ is the golden mean, and we say that $x > y$ if $x - y$ is positive. In essence, we are identifying $\mathbb{Z}^2$ with $\mathbb{Z} \oplus \tau\mathbb{Z} \subset \mathbb{R}$.

EXERCISE 3.2. *Show that $\left(\begin{smallmatrix} n_1 \\ n_2 \end{smallmatrix}\right)$ is positive if and only if $\left(\begin{smallmatrix} 0 & 1 \\ 1 & 1 \end{smallmatrix}\right)\left(\begin{smallmatrix} n_1 \\ n_2 \end{smallmatrix}\right)$ is positive. That is, the order structure is a property of the direct limit, and not just of each individual G_α.*

EXERCISE 3.3. *Show that $\left(\begin{smallmatrix} n_1 \\ n_2 \end{smallmatrix}\right)$ is positive if and only if, for m sufficiently large, both entries of $\left(\begin{smallmatrix} 0 & 1 \\ 1 & 1 \end{smallmatrix}\right)^m \left(\begin{smallmatrix} n_1 \\ n_2 \end{smallmatrix}\right)$ are positive.*

3.2. Cohomology and limits

Suppose we wish to compute the cohomology of a simple manifold, like S^2. We can view S^2 as a simplicial complex, as in Figure 3.1, with four vertices a, b, c, d, six edges $A = ab, B = ad, C = ac, D = db, E = bc, F = cd$ and four faces $\alpha = bcd, \beta = acd, \gamma = adb, \delta = abc$. These form a basis for our spaces of chains $C_0 = \mathbb{Z}^4$, $C_1 = \mathbb{Z}^6$ and $C_2 = \mathbb{Z}^4$, and we have boundary maps $\partial_2 : C_2 \to C_1$ and $\partial_1 : C_1 \to C_0$ given by the matrices

$$\partial_1 = \begin{pmatrix} -1 & -1 & -1 & 0 & 0 & 0 \\ 1 & 0 & 0 & 1 & -1 & 0 \\ 0 & 1 & 0 & 0 & 1 & -1 \\ 0 & 0 & 1 & -1 & 0 & -1 \end{pmatrix}, \tag{3.3}$$

$$\partial_2 = \begin{pmatrix} 0 & 0 & -1 & 1 \\ 0 & 1 & 0 & -1 \\ 0 & -1 & 1 & 0 \\ -1 & 0 & 1 & 0 \\ -1 & 0 & 0 & 1 \\ -1 & 1 & 0 & 0 \end{pmatrix}, \tag{3.4}$$

encoding the fact that $\partial A = b - a$, $\partial \alpha = -D - E - F$, etc. We then define cochain groups $C^n = (C_n)^*$, with basis dual to the basis for C_n, and have maps

$$0 \to C^0 \xrightarrow{\delta_0} C^1 \xrightarrow{\delta_1} C^2 \to 0, \tag{3.5}$$

where

$$\delta_0 = \partial_1^T = \begin{pmatrix} -1 & 1 & 0 & 0 \\ -1 & 0 & 1 & 0 \\ -1 & 0 & 0 & 1 \\ 0 & 1 & 0 & -1 \\ 0 & -1 & 1 & 0 \\ 0 & 0 & -1 & 1 \end{pmatrix}, \tag{3.6}$$

$$\delta_1 = \partial_2^T = \begin{pmatrix} 0 & 0 & 0 & -1 & -1 & -1 \\ 0 & 1 & -1 & 0 & 0 & 1 \\ -1 & 0 & 1 & 1 & 0 & 0 \\ 1 & -1 & 0 & 0 & 1 & 0 \end{pmatrix}. \tag{3.7}$$

We than have

$$H^0(S^2) \qquad = \ker(\delta_0) \qquad = \mathbb{Z}, \tag{3.8}$$

$$H^1(S^2) \quad = \ker(\delta_1)/Im(\delta_1) \quad = 0, \tag{3.9}$$

$$H^2(S^2) \qquad = C^2/Im(\delta_1) \qquad = \mathbb{Z}, \tag{3.10}$$

where the generator for H^0 is $(1, 1, 1, 1)^T$. This is the constant function that evaluates to one on each vertex. The generator for H^2 can be taken to be $(1,0,0,0)^T$, the function that evaluates to one on α and zero on the other faces. It could just as well be taken to be $(0,1,0,0)^T$ or $(0,0,1,0)^T$ or $(0,0,0,1)^T$. These are all different cochains, but they differ by elements of $Im(\delta_1)$ and represent the same class in H^2

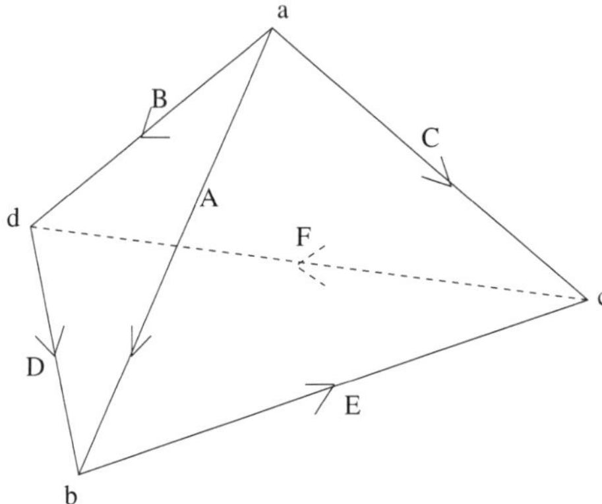

FIGURE 3.1. A sphere can be viewed as a tetrahedron.

This is a straightforward computation of the cohomology of S^2, but why did we use the particular simplicial structure of Figure 3.1? We could just as well have viewed the sphere as an octahedron, or an icosahedron, or as a cube with each face cut in half along a diagonal. How do we know that the cohomology of S^2 is a topological invariant and doesn't depend on the choice of simplicial structure?

To avoid any favoritism, we can consider *all* simplicial decompositions (i.e., triangulations) of S^2. If Σ_α and Σ_β are triangulations of S^2, we say that Σ_β is a refinement of Σ_α if every simplex (vertex, edge or face) in Σ_α is a union of simplices of Σ_β. In this case we write $\alpha < \beta$. It is not hard to show that any two triangulations have a common refinement, so the set of triangulations is a directed set.

If $\alpha < \beta$, then there is a simplicial map $i_{\beta\alpha} : \Sigma_\beta \to \Sigma_\alpha$, defined recursively as follows. (See Figure 3.2.) First, the vertices of Σ_α are mapped to themselves. Then the vertices of Σ_β that sit on edges of Σ_α are mapped to the endpoints of those edges is such a way that one portion of each edge expands to fill the entire edge, while the remaining portions are collapsed to points. This step is *not* canonically defined, and involves choices. Then the vertices that sit inside the faces of Σ_α are mapped to the vertices of those faces such that one portion of the face expands to fill the face and the remaining portions are collapsed onto edges. The map $i_{\beta\alpha}$ then induces a map on chains $C_k(\Sigma_\beta) \to C_k(\Sigma_\alpha)$, a transpose map on cochains $C^k(\Sigma_\alpha) \to C^k(\Sigma_\beta)$ and a map on cohomology $\rho_{\beta\alpha} : H^k_{\text{simplicial}}(\Sigma_\alpha) \to H^k_{\text{simplicial}}(\Sigma_\beta)$. Although the map $i_{\beta\alpha}$ involved some arbitrary choices, the induced map $\rho_{\beta\alpha}$ is canonically defined. (For a proof, see [**BT**]. Although they do not prove this result exactly as stated, their proof of Lemma 10.4.2 is easily modified.)

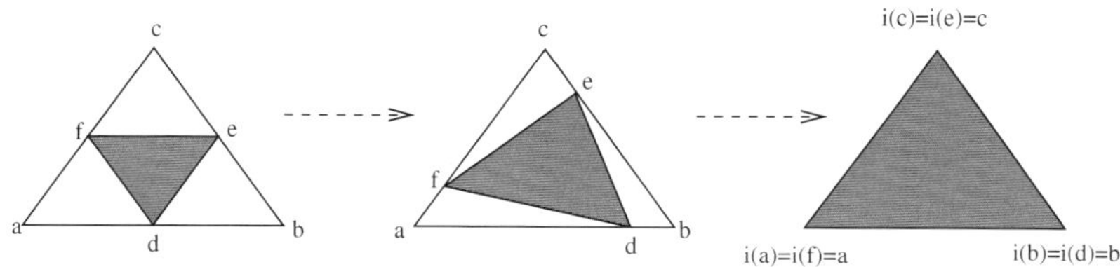

FIGURE 3.2. A simplicial map sending one cell in the subdivision of a triangle to the whole triangle, while collapsing the other cells.

We can now define the S-cohomology

$$H_S^k(S^2) = \varinjlim H_{\text{simplicial}}^k(\Sigma_\alpha), \tag{3.11}$$

where the direct limit is taken over all triangulations Σ_α using the maps ρ. This is manifestly a topological invariant.

This approach is not usually presented in textbooks. More commonly, it is proven that $\rho_{\beta\alpha}$ is an isomorphism, so all simplicial complexes given the same answer, so we don't need a direct limit. However, that implies that the common answer *is* the direct limit, so

$$H_S^k(S^2) = H_{\text{simplicial}}^k(\Sigma_\alpha) \tag{3.12}$$

for *any* triangulation of S^2. Another approach is to introduce singular cohomology, which is manifestly a topological invariant, and then prove that the singular cohomology of a simplicial complex equals its simplicial cohomology. The upshot is that simplicial cohomology, singular cohomology and S-cohomology are all the same on S^2, and more generally on spaces that admit a simplicial structure.

3.3. Čech cohomology of an open cover

Let X be a topological space and let $\mathcal{U} = \{U_\alpha\}$ be an open cover. The *nerve* of $\mathcal{U}$, denoted $N(\mathcal{U})$, is a simplicial complex with

(1) A vertex α for each non-empty open set U_α,
(2) An edge $\alpha\beta$ for each nonempty intersection $U_\alpha \cap U_\beta$,
(3) A face $\alpha\beta\gamma$ for each nonempty intersection $U_\alpha \cap U_\beta \cap U_\gamma$,
(4) An n-simplex for every non-empty intersection of $n+1$ open sets.

DEFINITION. *The Čech cohomology of the cover $\mathcal{U}$, denoted $\check{H}^*(\mathcal{U})$, is the simplicial cohomology of $N(\mathcal{U})$.*

Čech cohomology can be generalized to take values in any presheaf defined over $\mathcal{U}$, but we will only use the basic definition.

Note that the open cover does matter, as the following computations with $X = \mathbb{R}/\mathbb{Z} = S^1$ show. See Figure 3.3.

(1) X has an open cover consisting of a single set $U_0 = X$. In this case, $N(\mathcal{U})$ is a single point, and $\check{H}^0(\mathcal{U}) = \mathbb{Z}$ is the only non-trivial cohomology group.

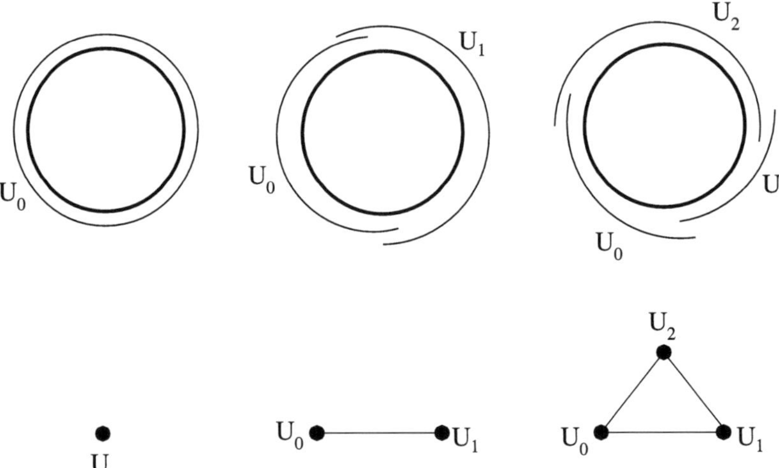

FIGURE 3.3. Three open covers of S^1 and their nerves

(2) We can cover the circle with two open sets, in which case the nerve
    is an interval. Again, $\check{H}^0(\mathcal{U}) = \mathbb{Z}$ is the only non-trivial cohomology
    group.
(3) If we cover the circle with three open sets as in Figure 3.3, then the
    nerve is a triangle, and we have $\check{H}^0(\mathcal{U}) = \check{H}^1(\mathcal{U}) = \mathbb{Z}$.

An open cover $\mathcal{V}$ is a *refinement* of $\mathcal{U}$ if every open set in $\mathcal{V}$ is contained in
an open set of $\mathcal{U}$. In this case, there is a simpicial map $i_{\mathcal{V}\mathcal{U}} : N(\mathcal{V}) \to N(\mathcal{U})$
that induces a map $\rho_{\mathcal{V}\mathcal{U}} : \check{H}^k(\mathcal{U}) \to \check{H}^k(\mathcal{V})$. See [**BT**], section 10, for details.
As with refinements of simplicial structures, the map $i_{\mathcal{V}\mathcal{U}}$ involves a number
of arbitrary choices, but the map $\rho_{\mathcal{V}\mathcal{U}}$ is independent of these choices.

DEFINITION. *The Čech cohomology of X, denoted $\check{H}^k(X)$, is the direct
limit of $\check{H}^k(\mathcal{U})$, where the limit is taken over all open covers of X.*

3.4. Cofinal sets, good covers and open stars

The definition of $\check{H}^k(X)$ is in the same spirit as the definitions of S-
cohomology and singular cohomology and looks hopelessly complicated at
first. How can we ever consider *all* open covers? Fortunately, it is often
possible to limit our attention to a subset of the open covers. If (as happens
frequently) ρ turns out to be an isomorphism on the Čech cohomology of
these covers, then the direct limit is the Čech cohomology of any one of
these covers.

If I is a directed set and $J \subset I$, we say that J is *cofinal* if for every $i \in I$
there exists a $j \in J$ with $i < j$.

THEOREM 3.1. *Let I be a directed set and J be a cofinal subset. If we
have groups G_i and homomorphisms ρ satisfying the conditions for a direct*

limit, then the direct limit of G_i over all $i \in I$ equals the direct limit of G_j over all $j \in J$.

PROOF. If $x \in G_i$, then x is identified with an element of G_j for some $j \in J$. Furthermore, if two elements $x \in G_{i_1}$ and $y \in G_{i_2}$ are identified in the direct limit under I, then they are identified in some G_{i_3} for some i_3 with $i_1 < i_3$, $i_2 < i_3$. But then they are identified in some G_j with $i_3 < j$. Since every generator of the direct limit of the G_i's is found in the direct limit of the G_j's, and since every relation among them is preserved, the two direct limits are the same. $\square$

For computing Čech cohomology, therefore, we do not need to consider all open covers. We just need to construct a cofinal set of open covers.

An open cover $\mathcal{U}$ of X is called a *good cover* if every open set in the cover is contractible, and every non-empty intersection of a finite number of open sets is contractible. In Figure 3.3, the first open cover of S^1 was not good, as U_1 was not contractible. The second was not good, as $U_1 \cap U_2$ was disconnected and hence not contractible. The third was good, and the Čech cohomology of that cover was isomorphic to the singular cohomology of S^1. This is not a coincidence!

Smooth manifolds and smooth branched manifolds always admit good covers. Give the manifold a Riemannian metric and take each open set to be a small metric ball around a point. Small balls are convex, and the intersection of convex sets is convex, and convex sets are contractible. This construction not only shows that good covers exist, but also that they are cofinal. If $\mathcal{U}$ is an arbitrary open cover, then every point in X has a ball that sits inside one of the open sets of $\mathcal{U}$. This collection of small balls is a good cover that refines $\mathcal{U}$.

This argument is not limited to (branched) manifolds. It applies just as well to finite CW complexes. All that is needed is a notion of "straight line" for which small balls around points are convex. However, the argument does *not* apply to tiling spaces. A neighborhood of a point in a tiling space is not a convex ball; it is a disk crossed with a Cantor set!

The key fact about good covers is this theorem, whose proof may be found in [**BT**]:

THEOREM 3.2. *If $\mathcal{U}$ and $\mathcal{V}$ are good covers of X with $\mathcal{V}$ refining $\mathcal{U}$, then $\rho_{\mathcal{V}\mathcal{U}} : \check{H}^k(\mathcal{U}) \to \check{H}^k(\mathcal{V})$ is an isomorphism.*

COROLLARY 3.3. *If X is a manifold or a finite CW complex, and if $\mathcal{U}$ is a good cover of X, then $\check{H}(X) = \check{H}(\mathcal{U})$.*

We have seen that Čech cohomology is computable, but how does it relate to other cohomology theories?

THEOREM 3.4. *If X is a finite simplicial complex, then $\check{H}^k(X)$ is isomorphic to the simplicial (and therefore singular) cohomology of X.*

PROOF. We construct a good open cover of X whose nerve is X itself. For each vertex a of X, let U_a be the union of a with the interiors of all of the simplices that have a as a vertex. This is called the *open star* of a. Since every point in X is either a vertex or is in the interior of a simplex, the sets $\{U_a\}$ cover X. U_a is star-shaped, hence contractible. Note that $U_a \cap U_b$ is non-empty if and only if a and b are endpoints of an edge e, in which case $U_a \cap U_b$ is the union of e and the interiors of all the simplices of dimension > 1 that have e as an edge, and is contractible. Likewise, $U_a \cap U_b \cap U_c$ is non-empty if and only if a, b, c are the vertices of a face f, in which case $U_a \cap U_b \cap U_C$ is the union of f and all the simplices of dimension > 2 that have f as a face. In other words, $N(\mathcal{U}) = X$. Combining this with Corollary 3.3 and the definition of Čech cohomology we have

$$\check{H}^k(X) = \check{H}^k(\mathcal{U}) = H^k_{\text{simplicial}}(N(\mathcal{U})) = H^k_{\text{simplicial}}(X). \tag{3.13}$$

$\square$

3.5. Cohomology of inverse limits

THEOREM 3.5. *Let X be the inverse limit of a sequence of spaces Γ_n induced by the maps $f_n : \Gamma_n \to \Gamma_{n+1}$. If each Γ_n is a space where good covers are cofinal (e.g., if Γ_n is a manifold or a finite CW complex), then*

$$\check{H}^k(X) = \varinjlim H^k(\Gamma_n). \tag{3.14}$$

PROOF. X is a subset of the product space $\Pi \Gamma_n$, and we have the natural projections $\pi_n : X \to \Gamma_n$. Since $\Pi \Gamma_n$ has the product topology, a basis for the open sets in X is given by pullbacks of open sets in the various Γ_n's. Since the good covers are cofinal on Γ_n, the pullbacks of good open covers are cofinal on X.

Let $\mathcal{U}_0$ be a good open cover of Γ_0, let $\mathcal{U}_1$ be a good open cover of Γ_1 that refines $f_1^{-1}(\mathcal{U}_0)$, and generally let $\mathcal{U}_n$ be a good open cover of Γ_n that refines $f_{n-1}^{-1}(\mathcal{U}_{n-1})$. Finally, let $\mathcal{V}_n = \pi_n^{-1}(\mathcal{U}_n)$. This is an open cover of X, but is generally *not* a good cover.

EXERCISE 3.4. *Show that $N(\mathcal{V}_n) = N(\mathcal{U}_n)$.*

EXERCISE 3.5. *Show that, by choosing the covers $\mathcal{U}_n$ appropriately, we can make the covers $\mathcal{V}_n$ cofinal among open covers of X. You will need to use the facts that good covers are cofinal on each Γ_n and link this with the $n \to \infty$ limit using a Cantor diagonalization argument.*

If the covers $\mathcal{V}_n$ are cofinal among covers of X (as in the preceding exercise), then $\check{H}^k(X) = \varinjlim \check{H}^k(\mathcal{V}_n)$. Now $\check{H}^k(\mathcal{V}_n)$ is the simplicial homology of $N(\mathcal{V}_n) = N(\mathcal{U}_n)$, and the simplicial cohomology of $N(\mathcal{U}_n)$ is $\check{H}^k(\mathcal{U}_n)$. However, $\mathcal{U}_n$ is a good cover of Γ_n, so $\check{H}^k(\mathcal{U}_n) = H^k(\Gamma_n)$. $\square$

Examples.

(1) Let X be the dyadic solenoid, which is the inverse limit of $\Gamma_n = S^1 = \mathbb{R}/\mathbb{Z}$ under the map $f_n(x) = 2x$. $H^0(S^1) = H^1(S^1) = \mathbb{Z}$, and $f_n^* : H^k(S^1) \to H^k(S^1)$ is the identity on H^0 and multiplication by 2 on H^1. $H^0(X)$ is then the direct limit of $\mathbb{Z}$ under the identity map and $\check{H}^1(X)$ is the direct limit of $\mathbb{Z}$ under multiplication by 2. That is, $\check{H}^0(X) = \mathbb{Z}$ and $\check{H}^1(X) = \mathbb{Z}[1/2]$.

This example illustrates the fact that inverse limit spaces typically do not have good covers. The dyadic solenoid is compact, so if it had a good cover it would have a finite good cover. In that case, $\check{H}^k(X)$ would be the simplicial cohomology of a finite simplicial complex. In particular, $\check{H}^1$ would be finitely generated as an Abelian group. However, $\mathbb{Z}[1/2]$ is not finitely generated.

(2) By the Anderson-Putnam construction, the Fibonacci tiling space is the inverse limit of a fixed approximant (call it Γ) shown in Figure 3.4. This approximant has $H^0 = \mathbb{Z}$ and $H^1 = \mathbb{Z}^2$. We restrict attention to H^1, since σ^* is the identity on H^0, as it is with all substitution tiling spaces.

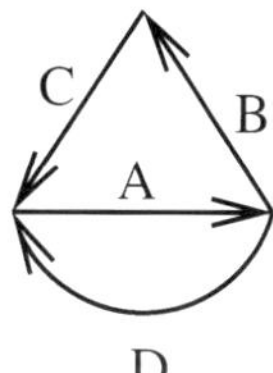

FIGURE 3.4. The (collared) Anderson-Putnam complex for the Fibonacci tiling

EXERCISE 3.6. *Show that the duals to the D and B edges give a basis for $H^1(\Gamma)$.*

EXERCISE 3.7. *Show that the substitution σ induces a map $\sigma^* : H^1(\Gamma) \to H^1(\Gamma)$ whose matrix is $\left(\begin{smallmatrix} 0 & 1 \\ 1 & 1 \end{smallmatrix}\right)$.*

From these exercises, we conclude that $\check{H}^1$ of the Fibonacci tiling space is $\mathbb{Z} \oplus \tau\mathbb{Z}$.

(3) Consider the Thue-Morse substitution $a \to ab$, $b \to ba$. All three-letter words are allowed except aaa and bbb. In other words, there are six collared tiles in the Thue-Morse tiling space.

EXERCISE 3.8. *Construct the Anderson-Putnam complex using uncollared tiles. Compute H^1 of this complex and the action of σ^* on H^1. From that, compute the cohomology of the inverse limit space.*

EXERCISE 3.9. *Construct the Anderson-Putnam complex using* **collared** *tiles. Compute H^1 of this complex and action of σ^*. From that, compute $\check{H}^1$ of the Thue-Morse tiling space.*

Your answers to the last two exercises should be different. The uncollared inverse limit has $\check{H}^1 = \mathbb{Z}[1/2]$, while the collared inverse limit has $\check{H}^1 = \mathbb{Z}[1/2] \oplus \mathbb{Z}$. This shows that the inverse limit of the uncollared Anderson-Putnam complex is not homeomorphic to the Thue-Morse tiling space. Since the Thue-Morse substitution doesn't force the border, we must use collared tiles!

(4) The half-hex tiling forces the border, so we do not have to use collared tiles. The (uncollared) Anderson-Putnam complex is shown in Figure 3.5, where all parallel short edges are identified. The approximant Γ has $H^0(\Gamma) = \mathbb{Z}$, $H^1(\Gamma) = \mathbb{Z}^2$ and $H^2(\Gamma) = \mathbb{Z}^3$. The pullback σ^* of the substitution acts trivially on H^0, so $\check{H}^0(\Omega_\sigma) = \mathbb{Z}$. The map σ^* multiplies all elements of H^1 by 2, so $\check{H}^1(\Omega_\sigma) = \mathbb{Z}[1/2]^2$. H^2 is a little more complicated. The substitution acts by the matrix $M = \begin{pmatrix} 2 & 1 & 1 \\ 1 & 2 & 1 \\ 1 & 1 & 2 \end{pmatrix}$, whose eigenvalues are 4, 1 and 1, with eigenvectors $\begin{pmatrix} 1 \\ 1 \\ 1 \end{pmatrix}$, $\begin{pmatrix} 1 \\ -1 \\ 0 \end{pmatrix}$ and $\begin{pmatrix} 0 \\ 1 \\ -1 \end{pmatrix}$. The direct limit of the direct sum of the eigenspaces is $\mathbb{Z}[1/4] \oplus \mathbb{Z}^2$. (As a set, $\mathbb{Z}[1/4]$ is the same as $\mathbb{Z}[1/2]$, but we write the $\mathbb{Z}[1/4]$ for $\check{H}^2$ and $\mathbb{Z}[1/2]$ for $\check{H}^1$ to emphasize how the different elements scale with substitution.) However, the direct sum of the three eigenspaces is *not* all of $\mathbb{Z}^3$! The entries of each eigenvector add up to a multiple of 3, so there is no way to write $(1, 0, 0)$ as a sum of eigenvectors. The direct sum of the eigenspaces is an index-3 subgroup of $\mathbb{Z}^3$, and hence that an index-3 subgroup of $\check{H}^2(\Omega_\sigma)$ is isomorphic to $\mathbb{Z}[1/4] \oplus \mathbb{Z}^2$. It turns out that $\check{H}^2(\Omega_\sigma)$ is *also* isomorphic to $\mathbb{Z}[1/4] \oplus \mathbb{Z}^2$, but that isomorphism is more subtle than just looking at the eigenspaces.

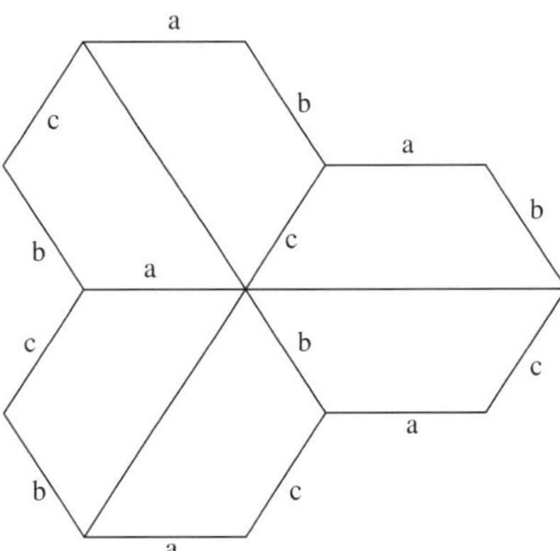

FIGURE 3.5. The Anderson-Putnam complex for the half-hex tiling

(5) The Anderson-Putnam complex for the Penrose substitution has $H^0 = \mathbb{Z}$, $H^1 = \mathbb{Z}^5$ and $H^2 = \mathbb{Z}^8$. The substitution acts on each of these by an invertible matrix, so the tiling space also has $\check{H}^0 = \mathbb{Z}$, $\check{H}^1 = \mathbb{Z}^5$ and $\check{H}^2 = \mathbb{Z}^8$.

(6) When a substitution in 2 or more dimensions does not force the border, the collared Anderson-Putnam complex tends to be too complicated to analyze by hand. However, many of the calculations can be done by computer. See [**Gah**].

3.6. Shape deformations

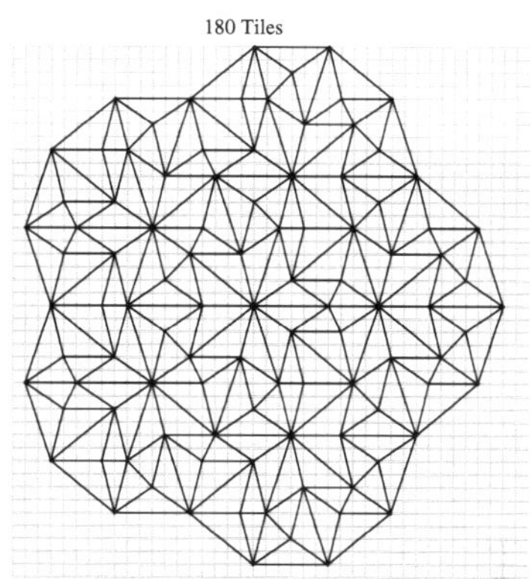

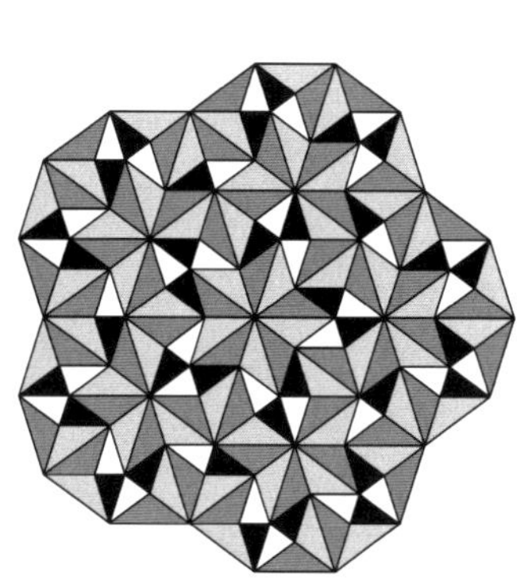

FIGURE 3.6. Ordinary and rational Penrose tilings.

Consider the tilings in Figure 3.6. The first (call it T) is a patch of a Penrose tiling. The second (call it T') is combinatorially identical to the first, but the shapes of the forty different prototiles have been changed. You can see that the rotational symmetry of the first tiling has been broken, and that tiles pointing in some directions are fatter than tiles pointing in other directions. The shapes of the tiles of T' are chosen so that all of the vertices of T' have integer coordinates, and T' is called a *rational Penrose tiling*.

Should we consider T and T' to be the same tiling or not? Are they MLD? Are their hulls topologically conjugate? Are their hulls homeomorphic? These questions can be answered with tiling cohomology.

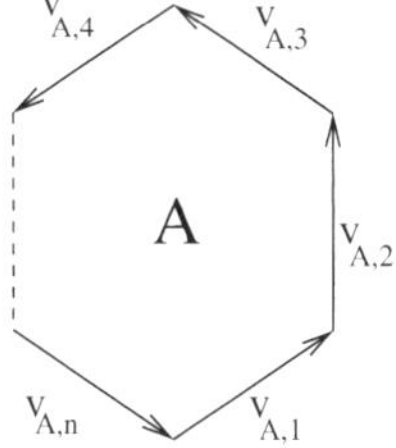

FIGURE 3.7. The shape of an n-gon is determined by the n vectors that describe its edges.

Parametrizing Shapes. If A is an n-gon, then the shape of A is described by a collection of n vectors $\{v_{A,1}, \ldots, v_{A,n}\}$, in the plane, with the vector $v_{A,j}$ describing the displacement along the j-th edge of A. These vectors cannot be chosen arbitrarily. They are subject to the condition

$$\sum_j v_{A,j} = 0, \tag{3.15}$$

since going all the way around the polygon has to leave you where you started! They are also subject to the condition that the boundary of A should be a positively oriented closed path that does not intersect itself. This is an open condition, meaning that any small change to the shape with $\sum_j v_{A,j} = 0$ will preserve the other conditions. In particular, infinitesimal changes to the shape are limited only by condition (3.15).

In a tiling, we cannot change the shapes of the different tiles independently. If two tiles share an edge, then the vector describing that edge must be the same for both tiles! At least infinitesimally, the shapes in a tiling are described by assigning a vector to each edge in a complex obtained by taking one copy of each tile type and identifying edges that meet somewhere in the tiling. But that's the (uncollared) Anderson-Putnam space of the tiling! In other words, shapes of tilings are described by vector-valued 1-cochains on the 0-th Gähler approximant Γ_0.

We could also consider shape deformations where tiles are deformed according to the pattern of their neighbors. That would be a vector-valued 1-cochain on the first Gähler approximant Γ_1. In general, we consider vector-valued 1-cochains of Γ_n with n arbitrary but finite.

Let f be such a cochain. f is closed, since for any tile A,

$$\delta f(A) = f(\partial A) = \sum_j v_{A,j} = 0. \tag{3.16}$$

That is the only local constraint on f:

THEOREM 3.6. [**SW**] *If T is a tiling with shape described by the cochain f, and if g is closed and sufficiently close to f, then there exists a tiling T' whose shapes are described by g and whose combinatorics are identical to those of T. The map $T \to T'$ can be extended to a homeomorphism $\Omega_T \to \Omega_{T'}$.*

If $g - f$ is the coboundary of a vector-valued 0-cochain α in some approximant Γ_n, then T' is locally derivable from T: just move each vertex a by $\alpha(a)$. Likewise, T is locally derivable from T'. However, the converse is also true: If $f - g$ is small and is not exact (in any of the approximants Γ_n), then T' is not locally derivable from T. In other words,

THEOREM 3.7. [**CS**] *Infinitesimal deformations of a tiling T, up to MLD equivalence, are parametrized by the closed cochains on arbitrary Γ_n, modulo exact cochains on arbitrary Γ_n, i.e., by the direct limit of $H^1(\Gamma_n) \otimes \mathbb{R}^2$, i.e., by $\check{H}^1(\Omega_T) \otimes \mathbb{R}^2$.*

The word "infinitesimal" is important. If $f - g$ is large, then there may be MLD maps between Ω_T and $\Omega_{T'}$ that do not preserve the combinatorial structure. For instance, if T is a substitution tiling with stretching factor λ and if $f = \lambda g$, then subdivision (without stretching) sends Ω_T to $\Omega_{T'}$, but typically does not send T to T'.

3.7. Topological conjugacy

We have classified shape deformations up to MLD equivalence, but what about topological conjugacy? There is a subspace $\mathcal{N} \subset \check{H}^1(\Omega_T, \mathbb{R}^2)$ such that, if $f - g \in \mathcal{N}$, then Ω_T is topologically conjugate to $\Omega_{T'}$. Classes in $\mathcal{N}$ are called *asymptotically negligible*. Shape changes of this type were among the first examples of topological conjugacies between tiling spaces that are not MLD.

The difficulty is identifying which cohomology classes are asymptotically negligible. For a general tiling space this is a difficult problem. For substitution tilings, however, there is a simple answer. The substitution σ sends Ω_T to itself, and induces a map σ^* from $\check{H}^1(\Omega_T, \mathbb{R}^2)$ to itself.

THEOREM 3.8. [**CS**] $\mathcal{N}$ *is the direct sum of all of the (generalized) eigenspaces of σ^* (acting on $\check{H}^1(\Omega_T, \mathbb{R}^2)$) with eigenvalue strictly less than 1 in magnitude.*

We now revisit the Penrose and rational Penrose tilings of Figure 3.6. The Penrose tiling space has $\check{H}^1(\Omega) = \mathbb{Z}^5$, so $\check{H}^1(\Omega, \mathbb{R}^2) = \mathbb{R}^{10}$. The substitution map has three eigenvalues: the golden mean τ, $1 - \tau$, and -1. The eigenspace E_τ and $E_{1-\tau}$ are 4-dimensional, while E_{-1} is 2-dimensional.

The original Penrose tiling expands by a linear factor of τ under substitution, so its shape function is entirely in E_τ. Since linear transformations commute with substitution, any linear transformation of the original Penrose tiling would yield a shape function in E_τ. In fact, the 4 dimensions of E_τ can be understood as parametrizing $GL(2, \mathbb{R})$; any tiling whose shape function is in E_τ is MLD to a linear transformation applied to the original Penrose tiling.

The eigenspace $E_{1-\tau}$ is algebraically conjugate to E_τ, and is asymptotically negligible. The eigenspace E_{-1} is not asymptotically negligible. Shape functions with components in this space are easily recognized, as this eigenspace is the only one to break the 180 degree rotational symmetry of the shapes in the original tiling.

The rational Penrose tiling preserves the 180 degree rotational symmetry, so its shape function lies entirely in the E_τ and $E_{1-\tau}$ eigenspaces. Since the edge vectors are integers, the shape function has a nonzero component in both E_τ and $E_{1-\tau}$, and these components are related by algebraic conjugation. As a result, the rational Penrose tiling is not MLD to the original, and is not MLD to a linear transformation of the original. It is, however, topologically conjugate to a linear transformation of the original.

CHAPTER 4

Relaxing the rules I: Rotations

So far we have considered tilings and translations but not rotations. We have said that moving a tile around gives a tile of the same type, but rotating the tile does not. Two tilings are considered close if they agree on a large ball up to a small translation, but not up to a small rotation. This approach has a sound historical basis. Tiling dynamical systems are a generalization of the dynamics of spaces of sequences (and their higher-dimensional analogs, called "subshifts"), for which rotations don't make sense.

But aren't we missing something? It is possible to view the Penrose tiling as having 40 different tile types, but it is much more natural to speak of 4 tile types in 10 orientations. Then we can talk about rotating Penrose tilings by multiples of 36 degrees to get other Penrose tilings (or maybe the same ones — there are four Penrose tilings with 5-fold rotational symmetry). The rotation group $\mathbb{Z}_{10}$ acts on the space of Penrose tilings, and we can study how this group action is reflected in the topology of the tiling space. None of this has any meaning if we insist on counting 40 different tile types.

An even more extreme example is the pinwheel tiling, a patch of which is shown in Figure 4.1. This tilings, invented by John Conway and extensively studied by Charles Radin [**Rad**], has only two tile types, but the tiles appear pointing in an infinite number of different directions!

FIGURE 4.1. The author's bathroom floor

The pinwheel substitution is shown in Figure 4.2. The basic tiles are left- and right-handed 1-2-$\sqrt{5}$ right triangles. Note that the central tile of a

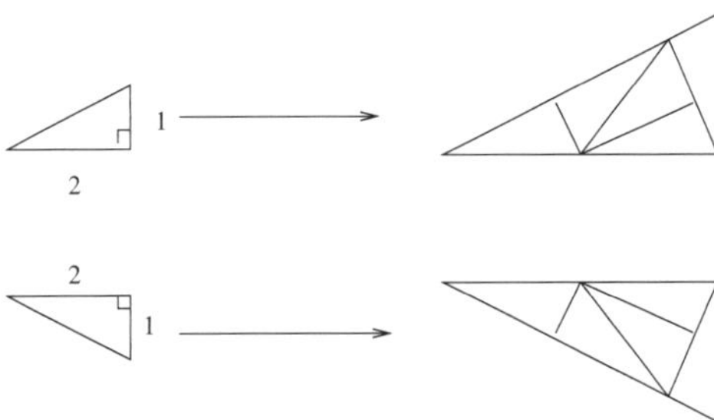

FIGURE 4.2. The pinwheel substitution

pinwheel supertile is rotated relative to the supertile by $\pm\tan^{-1}(1/2)$, which is an irrational angle.

Let $\alpha = \tan^{-1}(1/2)$, let r denote a positive rotation by α, and let s be rotation by 90 degrees. Taking the standard orientation for a triangle to be that with the long edge horizontal and the small angle to the left (as in Figure 4.2), the substitution matrix for the pinwheel can be expressed as

$$M = \begin{pmatrix} r + rs^2 & 2r^{-1} + r^{-1}s^3 \\ 2r + rs & r^{-1} + r^{-1}s^2 \end{pmatrix}. \tag{4.1}$$

That is, substituting a right-handed tile with orientation θ gives two right-handed tiles, with orientations $\theta+\alpha$ and $\theta+\alpha+\pi$ and three left-handed tiles, two with orientation $\theta + \alpha$ and one with orientation $\theta + \alpha + \pi/2$, as can be seen in Figure 4.2. Substituting a left-handed tile is similar, and is described by the second column of M. The effect of substituting n times is given by M^n, which has terms involving $r^n, r^{n-2}, \ldots, r^{-n}$. It is not hard to check that the resulting distribution of tile directions is uniformly distributed in $SO(2)$.

4.1. A new topology for tiling spaces

If we try to study the pinwheel tiling using the topology of chapter 1, we immediately run into trouble. If every direction counts as a different tile type, then there are an infinite number of tile types, and the pinwheel tiling space is not compact. To see this, take a sequence of tilings all of which have different tile types at the origin. This sequence has no convergent subsequence! Besides not being compact, the tiling space has no translation-invariant probability measures, precluding the use of ergodic theory. To make any sense of the pinwheel tiling spaces, we need a different topology.

Consider the following metric. Let tilings T and T' be considered ϵ-close if they agree on $B_{1/\epsilon}$, up to an ϵ-small *Euclidean* motion. That is, we allow translation by up to ϵ and rotation by up to ϵ.

For tilings with only finitely many tile directions, this yields the same topology as the metric of chapter 1, since rotations by arbitrarily small angles don't exist. For pinwheel tilings, it yields compactness. Any sequence of tilings has a subsequence where the tile type at the origin (meaning right-

or left-handed) converges, where the position of the tile at the origin converges, and where the direction of the tile at the origin converges. Since tiles meet full-edge to full-edge, there are only a finite number of ways to extend this to a tiling of B_R, where the radius R is arbitrary but fixed. We can therefore take a subsequence that converges on B_R. Repeating the process for increasing values of R, and applying the Cantor diagonalization trick, we get a subsequence that converges everywhere.

DEFINITION. *A tiling is* rotationally simple *if it satisfies three assumptions:*

(1) *There are only a finite number of tile types, up to Euclidean motion.*
(2) *Each tile is a polygon (or, in higher dimensions, a polytope).*
(3) *Tiles meet full-edge to full-edge.*

Note that to satisfy the third assumption we must consider the pinwheel tiles to be degenerate quadrilaterals, with an extra vertex in the middle of the long side. These assumptions are almost equivalent to *finite local complexity*, or FLC, the assumption that for any fixed R there can only be a finite number of patches of radius R, up to rigid motion. I say "almost" because these assumptions imply FLC, but a pattern like the Penrose chickens has FLC without satisfying these assumptions. However, by using the Voronoi cell trick, any tiling with FLC can be converted to a tiling satisfying these assumptions. Because of the rotational symmetry, some steps of the construction are surprisingly subtle; see [**Ran2**] for details.

EXERCISE 4.1. *Prove that a tiling satisfying assumptions 1–3 has FLC.*

4.2. Local structure and global topology

If T is a pinwheel tiling, what does a neighborhood of T in Ω_T look like? As before, we get the neighborhood by taking a large patch of T, wiggling it slightly, and extending it to infinity. The difference is that the small wiggle involves *three* degrees of freedom – two translations and a rotation. The discrete degrees of freedom are as before, and a neighborhood of T is homeomorphic to a 3-disk crossed with a Cantor set.

The global topology of Ω_T is similar to that of the hull of a simple tiling. Ω_T is connected but not path-connected. Indeed, it has uncountably many path components. Each path component is an orbit of the Euclidean group E_2, and is isomorphic to either E_2 or to the quotient of E_2 by a finite isotropy group.

4.3. Inverse limit structures

The hull of a simple tiling of $\mathbb{R}^2$ is the inverse limit of branched 2-dimensional manifolds. The hull of a pinwheel tiling can't possibly be the inverse limit of branched 2-manifolds, since its local structure involves three continuous degrees of freedom! Instead, we will construct Ω_T (and other pinwheel-like tiling spaces) as the inverse limit of branched 3-dimensional manifolds [**ORS**].

We can repeat the ideas of the Gähler and Anderson-Putnam constructions. In the Gähler construction, Γ_n is the set of instructions for placing an n-collared tile at the origin. This is equivalent to picking a point in an n-collared tile in standard orientation, plus specifying an angle. If the collared tile t lacks rotational symmetry, then the corresponding cell in Γ_n is $t \times S^1$. If t has k-fold rotational symmetry, then the cell is $(t \times S^1)/\mathbb{Z}_k$. In the Anderson-Putnam construction, we begin with $\tilde{\Gamma}_0 = \Gamma_1$, and each subsequent approximant is the same space, only rescaled by λ^n.

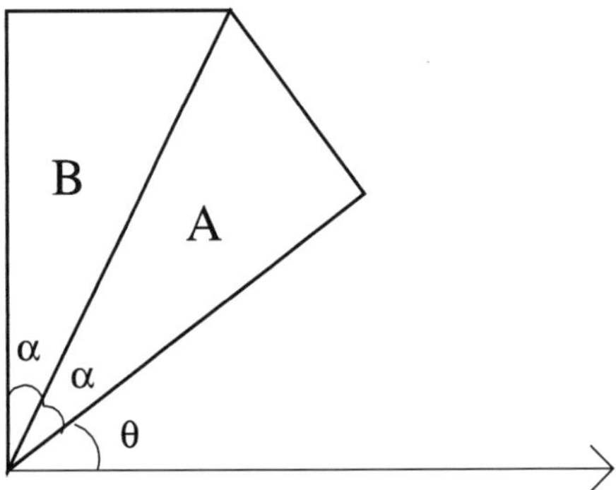

FIGURE 4.3. Two pinwheel tiles with a common hypotenuse

To stitch the tiles together, we must understand what happens when the origin sits on the boundary of two (or more) tiles. In Figure 4.3, the hypotenuse of the right-handed A tile is also the hypotenuse of the left-handed B tile. However, the A tile is rotated by θ relative to its standard orientation, while the B tile is rotated by $\theta + 2\alpha$. If p is a point on the hypotenuse of the A tile and q is the corresponding point on the B tile, then we identify (p, θ) with $(q, \theta + 2\alpha)$.

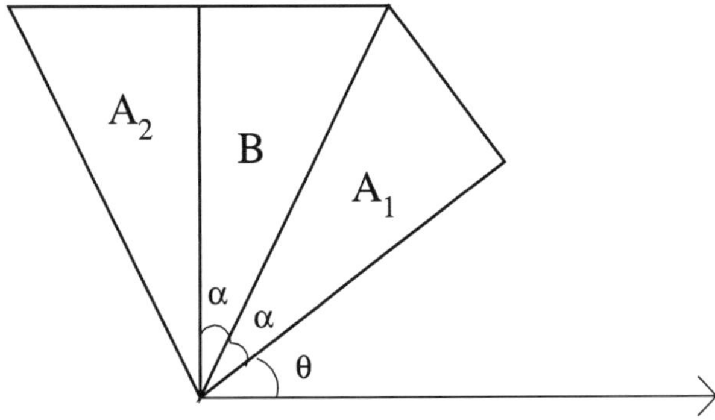

FIGURE 4.4. Three pinwheel tiles with a common vertex

Now consider the situation of Figure 4.4. If p, q, r are the small angle vertices of tiles A_1, B and A_2, then we identify $(p, \theta) \sim (q, \theta + 2\alpha) \sim (r, \theta + 2\alpha)$. This is no problem if we are using collared tiles, since then A_1 and A_2 have different tile types. However, if we tried to work with uncollared tiles, we would have to identify p with r, so $(p, \theta) \sim (p, \theta + 2\alpha)$. Likewise, $(p, \theta + 2\alpha) \sim (p, \theta + 4\alpha)$. Since α is irrational, we must identify an infinite number of different angles associated with p. Indeed, the entire circle corresponding to p collapses to a single point in Γ_0.

This is a disaster, as it means that Γ_0 is not a branched manifold at all! To avoid this problem, we *always* use collared tiles. The Gähler construction begins with Γ_1, not with Γ_0, and the Anderson-Putnam construction uses $\tilde{\Gamma}_0$.

EXERCISE 4.2. *Show that it is impossible for two identical collared tiles, rotated by an irrational angle relative to one another, to meet at a point that plays the same role in both tiles.*

There is a natural S^1-action on Γ_n that corresponds to rotation about the origin. The spaces Γ_n should be viewed as S^1-fibered spaces over their quotients by this S^1 action. The only subtlety is that the S^1 fibers may not all be the same size. If p is a point of k-fold rotational symmetry out to a large distance, as in Figure 4.5, then $(p, \theta) \sim (p, \theta + 2\pi/k)$. This does not imply that the space Γ_n is singular in any way. It just means that the quotient space Γ_n/S^1 has cone singularities.

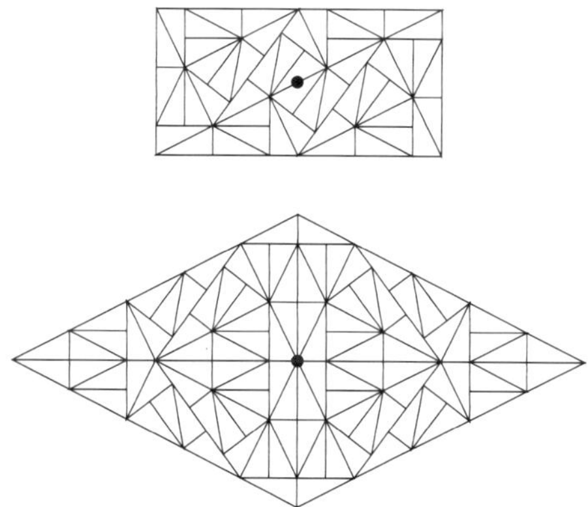

FIGURE 4.5. Two pinwheel patches with two-fold rotational symmetry

This is almost identical to the local structure of Seifert-fibered spaces. These are 3-manifolds with a locally free S^1 action. Some points are fixed by a finite subgroup $\mathbb{Z}_k$ of S^1, and the orbits of these points are smaller than the orbits of nearby points. Locally, the space looks like $M_\epsilon = (B_\epsilon \times S^1)/\mathbb{Z}_k$, where $(r, \theta, \phi) \sim (r, \theta + 2\pi/k, \phi - 2\pi/k)$. No points in $B_\epsilon \times S^1$ are fixed by the $\mathbb{Z}_k$ action, so the quotient M_ϵ is a smooth 3-manifold. Increasing ϕ while leaving (r, θ) alone gives a natural S^1 group action on M_ϵ. M_ϵ is *almost* an

S^1 bundle over $B_\epsilon/\mathbb{Z}_k$, but the fiber over the origin is only a circle of length $2\pi/k$, while the fibers over every other point have length 2π.

4.4. Quotients of the tiling space

In cohomology calculations, the S^1 factor mostly comes along for the ride, and we are reduced to studying the quotient space

$$\Omega_0 = \Omega_T/S^1 = \varprojlim(\Gamma/S^1, f). \tag{4.2}$$

Each approximant Γ_n/S^1 is a branched 2-orbifold. That is, it looks like a branched 2-manifold everywhere except at a finite number of cone points. These cone points have little impact on the rational cohomologies of Ω_0 and Ω_T.

THEOREM 4.1. *The rational cohomology of Ω_T can be computed from that of Ω_0, and vice versa. Specifically, $H^k(\Omega_T, \mathbb{R}) = H^k(\Omega_0 \times S^1, \mathbb{R}) = H^k(\Omega_0, \mathbb{R}) \oplus H^{k-1}(\Omega_0, \mathbb{R})$.*

The proof [**BDHS**] involves spectral sequences and the fact that the S^1 bundle admits a closed global angular form.

The integer cohomology is subtler and often *does* detect the cone points, with torsion in $H^2(\Omega_T)$ coming from the exceptional fibers. Suppose that we have two different tilings with 2-fold rotational symmetry, as in Figure 4.5, and a third tiling with no rotational symmetry. The patterns of the symmetric tilings, rotated about the origin, correspond to circles a and b of length π in Γ_n, while the pattern of the asymmetric tiling corresponds to a third circle c of length 2π. The circle c is homologous to both $2a$ and $2b$, but a and b are not homologous to each other. As a result, $H_1(\Gamma_n)$ has a $\mathbb{Z}_2$ factor generated by $a - b$. By the universal coefficients theorem, this implies that there is a $\mathbb{Z}_2$ factor in $H^2(\Gamma_n)$. If the tilings have rotational symmetry out to infinity, then this torsion exists in $H^2(\Gamma_n)$ for every n and survives to the limiting $H^2(\Omega_T)$.

The moral of the story is that infinite rotations do not give us any new and interesting cohomology, but finite rotations do! It's natural, then, to apply our rotational techniques to tilings with only finite rotational symmetry, like the chair tilings (with $\mathbb{Z}_4$ symmetry) and the Penrose tilings ($\mathbb{Z}_{10}$).

4.5. Three spaces of chair tilings

There are three tiling spaces that we can make out of chair tilings. We have already met the first one, which we denote Ω_1, obtained by taking the closure of the *translational* orbit of any one chair tiling. We can get a bigger space Ω_{rot} by taking the tilings in Ω_1 and rotating them by arbitrary angles. Any one tiling in Ω_{rot} will only have tiles pointing in four directions, but the directions that appear in one tiling do not have to match the directions appearing in another. The finite rotation group $\mathbb{Z}_4$ acts on Ω_1, the infinite rotation group $SO(2) = S^1$ acts on Ω_{rot}, and either quotient yields the space Ω_0 of chair tilings modulo rotation about the origin.

To summarize,

$$\Omega_1 \;=\; \{\text{All chair tilings with edges parallel to the coordinate axes}\}$$
$$=\; \{\text{Closure of the translational orbit of one chair tiling}\},$$

$$\Omega_{rot} \;=\; \{\text{All chair tilings in any orientation}\}$$
$$=\; \{\text{Closure of the Euclidean orbit of any one chair tiling}\},$$

$$\Omega_0 \;=\; \{\text{Chair tilings modulo rotation}\}$$
$$=\; \Omega_1/\mathbb{Z}_4 = \Omega_{rot}/S^1.$$

We could likewise produce three types of Penrose tiling spaces and three types of half-hex tiling spaces, only with $\mathbb{Z}_{10}$ (for Penrose) or $\mathbb{Z}_6$ (for half-hex) instead of $\mathbb{Z}_4$. The construction applies to any simple tiling space, with $\mathbb{Z}_4$ replaced by a rotational symmetry group that acts naturally on Ω_1.

Rotations and Cohomology. Working with the chair tilings, let r denote rotation by $\pi/2$. Since r maps Ω_1 to itself, it induces a map $r^* : \check{H}^*(\Omega_1, \mathbb{R}) \to \check{H}^*(\Omega_1, \mathbb{R})$. We can then decompose $\check{H}^*(\Omega_1, \mathbb{R})$ into irreducible representations of $\mathbb{Z}_4$.

We can play the same game with integer cohomology. The map r induces a map r^* from $\check{H}^*(\Omega_1)$ to itself. We must be careful in decomposing integer cohomology groups into irreducible representations of a rotation group, since integer cohomology groups are not vector spaces! Strictly speaking, the action of r^* on $\check{H}^*(\Omega_1)$ means that $H^*(\Omega_1)$ is a module over the group ring of $\mathbb{Z}_4$ (that is, over $\mathbb{Z}[r]/(r^4 - 1)$). Not all such modules can be written as direct sums of modules that come from irreducible representations of the group.

To see how a problem *might* arise, let s be a generator of $\mathbb{Z}_2$, and let s act on $\mathbb{Z}^2$ by $s(a,b) = (b,a)$. Both irreducible representations of $\mathbb{Z}_2$ are one-dimensional, and in the nontrivial representation, s acts by multiplication by -1. In our module, s acts trivially on all multiples of the vector $(1,1)$, and acts by -1 on all multiples of $(1,-1)$, but $\mathbb{Z}^2$ is not the direct sum of those two sub-modules. There is no way to write $(1,0)$ as a linear combination of $(1,1)$ and $(1,-1)$ with integer coefficients! The direct sum of the sub-modules generated by $(1,1)$ and $(1,1)$ is an index-2 subgroup of $\mathbb{Z}^2$, namely those points (a,b) with $a+b$ even.

In general, if a tiling space has symmetry group $\mathbb{Z}_n$, we can always decompose a finite-index subgroup of $\check{H}^k(\Omega_1)$ into a direct sum of irreducible representations of $\mathbb{Z}_n$. In computing $\check{H}^k(\Omega_1)$, we can do all of our calculations one representation at a time, but at the end we have to consider a possible finite extension. This problem does not arise when computing rational cohomology, and (as we shall see) it does not arise with the chair tiling.

The rotation r also acts on Ω_{rot}, but here r is homotopic to the identity map, so r^* is trivial. As with pinwheel-like spaces, the cohomologies of Ω_{rot} and Ω_1 are closely related [**BDHS**]. These relations are summarized as follows:

$$
\begin{aligned}
\check{H}^*(\Omega_0) &= \text{rotationally invariant part of } \check{H}^*(\Omega_1) \\
\check{H}^*(\Omega_{rot}, \mathbb{R}) &= H^*(\Omega_0 \times S^1, \mathbb{R}), \text{ so} \\
\check{H}^k(\Omega_{rot}, \mathbb{R}) &= H^k(\Omega_0, \mathbb{R}) \oplus H^{k-1}(\Omega_0, \mathbb{R}), \text{ but} \\
\check{H}^2(\Omega_{rot}, \mathbb{Z}) &= \check{H}^2(\Omega_0, \mathbb{Z}) \oplus \check{H}^1(\Omega_0, \mathbb{Z}) \\
&\quad \oplus \text{ possible torsion from exceptional fibers.}
\end{aligned}
$$

The spaces Ω_{rot} and Ω_0 are geometrically natural, but the richest cohomology theory is that of Ω_1.

4.6. A flat-Earth calculation

To illustrate these ideas, we will compute the cohomology of the various tiling spaces that come from the arrow substitution of Figure 4.6. There is only one tile type, appearing in four orientations in any given tiling. Arrow tilings are MLD to chair tilings, with the correspondence shown in Figure 4.7. Each chair tile can be viewed as a triple of arrow tiles, and the tiles in an arrow tiling can be grouped into threes to form a chair tiling.

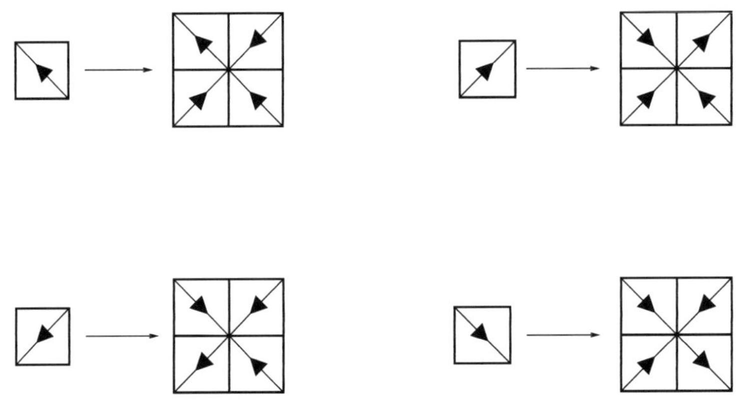

FIGURE 4.6. The arrow substitution.

Unfortunately, neither the arrow substitution nor the chair substitution forces the border. To do a proper calculation, we must use collared tiles. There are 13 collared arrow tiles, each in 4 orientations. Using a clever partial collaring scheme, this can be reduced to 6 collared tiles, each in 4 orientations. That's well within the limits of calculations by hand, and we will revisit this calculation in Chapter 6, but it is far too involved to serve as an Illustrative Example here.

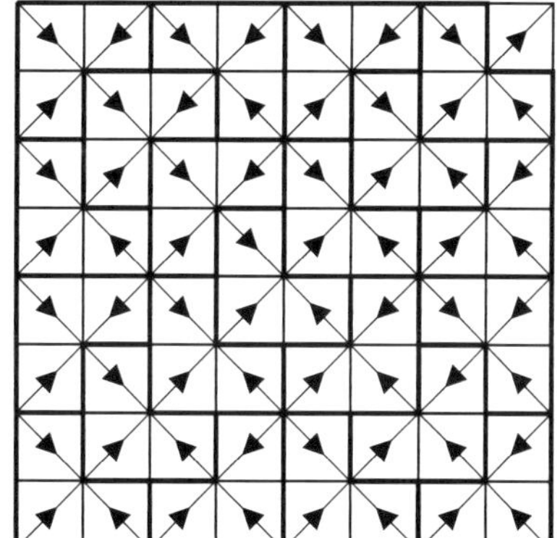
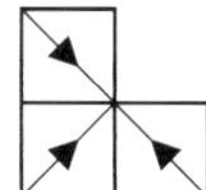

FIGURE 4.7. A chair tile equals three arrow tiles, and the tiles in an arrow tiling can be grouped into chairs.

Instead, we will emulate the Flat Earth Society and *pretend* that the arrow substitution forces the border. We know it doesn't, but we'll do our computations using the uncollared Anderson-Putnam complex anyway. By sheer luck, this procedure gives essentially the right answer. Although the natural map from the tiling space to the inverse limit of the Anderson-Putnam complex isn't a homeomorphism, the two spaces do have isomorphic cohomology groups.

We take the tile with an arrow pointing northeast to be the standard orientation. Call this tile A. The tiles with arrows pointing northwest, southwest and southeast, respectively, are then rA, r^2A and r^3A. We label the edges $(\alpha, \beta, \gamma, \delta)$ and vertices (a, b, c, d) of A as in Figure 4.8. Likewise, the edges of rA are $(r\alpha, r\beta, r\gamma, r\delta)$, etc. We then identify edges and vertices where two tiles meet.

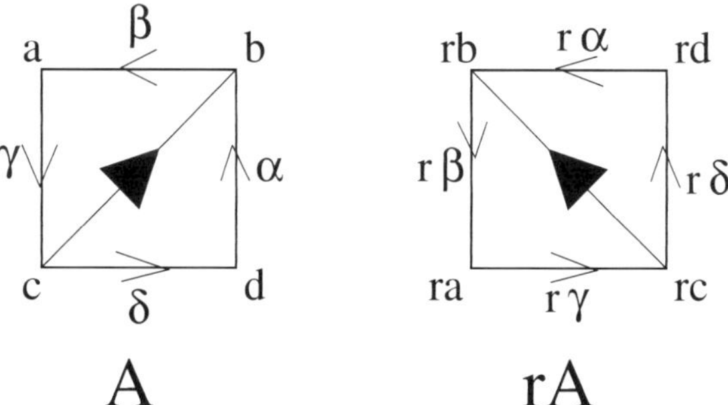

FIGURE 4.8. The edges and vertices of A and rA. The markings on r^2A and r^3A are similar.

Since rA can appear to the right of A, we identify α with $-r\beta$, meaning $r\beta$ with the opposite orientation. Likewise we identify b with rb, and d with ra. Since rA can also appear to the left of rA, we get that $\gamma = -r\delta$, $a = rd$, $c = rc$. Since rA can also appear below A, we get $\delta = -r\alpha$, $c = rb$, $d = rd$, and since rA can appear above A we get $\beta = -r\gamma$, $a = ra$, $b = rc$. By seeing how rA and r^2A can meet, we get the same relations multiplied by r (e.g., $r\alpha = -r^2\beta$). Seeing how r^2A and r^3A can meet, and how r^3A and A can meet, gives the above relations multiplied by r^2 and by r^3. The upshot is that

$$\alpha = -r\beta, \quad \beta = -r\gamma, \quad \gamma = -r\delta, \quad \delta = -r\alpha \tag{4.3}$$

and

$$a = d = ra = rd, \qquad b = c = rb = rc. \tag{4.4}$$

In other words, there is only one tile A, up to rotation, there is only one edge α, up to rotation, and there are two vertices, a and b, as shown in Figure 4.9. Note that the horizontal edges in Figure 4.9 point in the opposite direction as the corresponding edges in Figure 4.8, reflecting the fact that $\alpha = -r\beta$. The boundary map and remaining relations are:

$$\partial A = (1 - r + r^2 - r^3)\alpha, \tag{4.5}$$

$$\partial\alpha = b - a, \tag{4.6}$$

$$A = r^4A, \quad \alpha = r^4\alpha, \quad a = ra, \quad b = rb. \tag{4.7}$$

The substitution acts by

$$\sigma(A) = (2 + r + r^{-1})A, \quad \sigma(\alpha) = (1 - r^2)\alpha, \quad \sigma(a) = b, \quad \sigma(b) = b. \tag{4.8}$$

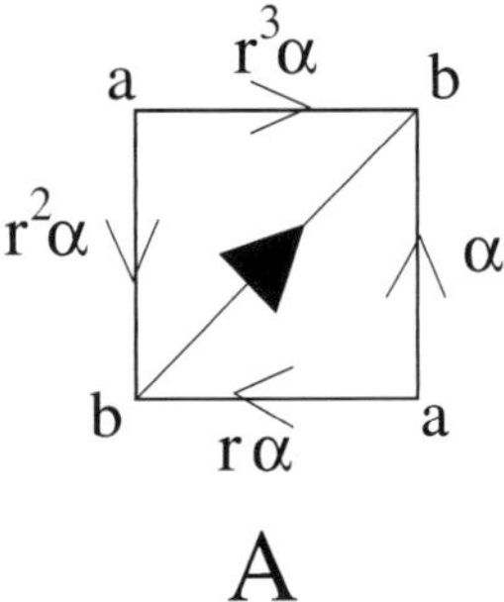

FIGURE 4.9. The edges and vertices of A after all identifications have been made.

There are three irreducible representations of $\mathbb{Z}_4$. The "scalar" ($r = 1$) and "pseudoscalar" ($r = -1$) representations are 1-dimensional, and there is a 2-dimensional "vector" representation with $r^2 = -1$. A and α have no nontrivial relations, and so appear in all representations. The vertices a and b appear only in the scalar representation.

Scalar representation: $r = 1$. Our chain groups are $C_0 = \mathbb{Z}^2$ (with generators a and b), $C_1 = \mathbb{Z}$ and $C_2 = \mathbb{Z}$. Since $1 - r + r^2 - r^3 = 0$, the boundary map from $\partial_2 : C_2 \to C_1$ is zero, while $\partial_1 : C_1 \to C_0$ is given by $\left(\begin{smallmatrix} -1 \\ 1 \end{smallmatrix} \right)$. Dualizing, we have $C^0 = \mathbb{Z}^2$, $C^1 = \mathbb{Z}$, $C^2 = \mathbb{Z}$, with $\partial_1^T = (-1, 1)$ and $\partial_2^T = 0$. Since ∂_1 is injective we have $H^1 = 0$, while $H^0 = H^2 = \mathbb{Z}$.

Now we look at the direct limit of the cohomologies. Since $2 + r + r^{-1} = 4$, substitution acts on H^2 by multiplication by 4, so the direct limit of H^2 is $\mathbb{Z}[1/4]$. Meanwhile, the direct limit of H^0 is $\mathbb{Z}$.

Pseudoscalar representation: $r = -1$. In the $r = -1$ representation, there are no vertices, and $C_1 = C_2 = \mathbb{Z}$, as before. Since $1 - r + r^2 - r^3 = 4$, we have $\partial_2 = 4$, and likewise $\partial_2^T : C^1 \to C^2$ is multiplication by 4, so H^0 and H^1 are trivial while $H^2 = \mathbb{Z}_4$. However, $2 + r + r^{-1} = 0$, so the substitution map $\sigma^* : H^2 \to H^2$ is trivial, and nothing survives to the direct limit.

Vector representation: $r^2 = -1$. Since this is a 2-dimensional representation, we have $C_1 = C_2 = \mathbb{Z}^2$, while $C_0 = 0$. Since $1 - r + r^2 - r^3 = 0$, there is no boundary map, and we have $H^1 = H^2 = \mathbb{Z}^2$. Since $1 - r^2 = 2$, the substitution map $H^1 \to H^1$ is multiplication by 2, and the direct limit is $\mathbb{Z}[1/2]^2$. Likewise, $2 + r + r^{-1} = 2$, so the substitution map $H^2 \to H^2$ is multiplication by 2, and the direct limit of H^2 is $\mathbb{Z}[1/2]^2$.

Putting it all together. Counting the contributions from all four representations, we get the Čech cohomology of Ω_1 for the arrow tiling:

$$\check{H}^0(\Omega_1) = \mathbb{Z} \tag{4.9}$$

scales trivially and rotates like a scalar. The generator 1 just counts points.

$$\check{H}^1(\Omega_1) = \mathbb{Z}[1/2]^2 \tag{4.10}$$

scales like length and rotates like a vector. The element $\left(\begin{smallmatrix} 2^{-n} \\ 0 \end{smallmatrix} \right)$ counts horizontal edges of order-n supertiles, while $\left(\begin{smallmatrix} 0 \\ 2^{-n} \end{smallmatrix} \right)$ counts vertical edges of order-n supertiles. The two pieces of

$$\check{H}^2(\Omega_1) = \mathbb{Z}[1/4] \oplus \mathbb{Z}[1/2]^2 \tag{4.11}$$

scale and rotate differently. The $\mathbb{Z}[1/4]$ piece scales like area and rotates like a scalar. The generator 4^{-n} counts order-n supertiles, making no distinction between the different types. The $\mathbb{Z}[1/2]^2$ piece scales like length, not like area, and rotates like a vector. The element $\left(\begin{smallmatrix} 2^{-n} \\ 0 \end{smallmatrix} \right)$ counts the number of order-n supertiles of type A minus the number of type $r^2 A$, while $\left(\begin{smallmatrix} 0 \\ 2^{-n} \end{smallmatrix} \right)$ counts the number of order-n supertiles of type rA minus the number of type $r^3 A$.

[Note: The direct sum of the irreducible representations of $\mathbb{Z}_4$ is not all of $\mathbb{Z}^4$, but there's no harm done. Although the basis elements are not in the direct sum, four times each basis element is. Since all elements of $\check{H}^1$ and $\check{H}^2$ are divisible by four anyway, the difference between $(1, 0, 0, 0)^T$ and $(4, 0, 0, 0)^T$ is irrelevant.]

Other spaces of arrow tilings. What about the cohomology of Ω_0 and Ω_{rot}? The Anderson-Putnam complex of Ω_0 is exactly Figure 4.9, with α identified with $r\alpha$, $r^2\alpha$ and $r^3\alpha$. The calculation of $\check{H}^*(\Omega_0)$ is *precisely* the scalar part of the $\check{H}^*(\Omega_1)$ calculation, with the result that

$$\check{H}^0(\Omega_0) = \mathbb{Z}, \qquad \check{H}^1(\Omega_0) = 0, \qquad \check{H}^2(\Omega_0) = \mathbb{Z}[1/4]. \tag{4.12}$$

As for Ω_{rot}, we know its cohomology is that of $\Omega_0 \times S^1$, plus torsion terms from tilings with symmetry. There is only one tiling with symmetry, namely the tiling with four infinite-order supertiles pointing outwards from the origin. Since it takes two exceptional fibers to generate torsion, there is no torsion. The upshot is that

$$\check{H}^0(\Omega_{rot}) = \check{H}^1(\Omega_{rot}) = \mathbb{Z}, \qquad \check{H}^2(\Omega_{rot}) = \check{H}^3(\Omega_{rot}) = \mathbb{Z}[1/4], \tag{4.13}$$

with the $\mathbb{Z}$ terms coming from $\check{H}^0(\Omega_0)$ and the $\mathbb{Z}[1/4]$ terms coming from $\check{H}^2(\Omega_0)$.

Back to the round Earth. Now that we have completed our "flat Earth" calculation, I should confess one point where the results are misleading. For the chair tiling, consider the class in H^2 that counts chair tiles. Since each chair tile has three arrow tiles, this is actually *one third* of the class that counts arrow tiles. However, $\frac{1}{3}$ is not in $\mathbb{Z}[1/4]$! Something must be wrong.

In fact, $\check{H}^1(\Omega_1) = \mathbb{Z}[1/2]^2$ and $\check{H}^2(\Omega_1)$ is isomorphic to $\mathbb{Z}[1/2]^2 \oplus \mathbb{Z}[1/4]$, exactly as above, but the class that counts arrow tiles is *not* a generator of the $\mathbb{Z}[1/4]$ subgroup of $\check{H}^2(\Omega_1)$. It is three times a generator of $\mathbb{Z}[1/4]$. If we let $1 \in \mathbb{Z}[1/4]$ denote the class that counts arrow tiles, then we should write $\check{H}^2(\Omega_1) = \frac{1}{3}\mathbb{Z}[1/4] \oplus \mathbb{Z}[1/2]^2$. The class $\frac{1}{3} \in \frac{1}{3}\mathbb{Z}[1/4]$ cannot be described in terms of uncollared arrow tiles, but is easily described in terms of collared arrow tiles or in terms of chair tiles.

4.7. Penrose cohomology

As a second example, we compute the cohomologies of the various spaces of Penrose tilings. Since the Penrose substitution forces the border and has tiles in only a finite number of orientation (per tiling), we do not need to collar. The substitution matrices are invertible in all representations, so we need only compute the cohomology of the first Anderson-Putnam approximant.

We have four tile types, each in 10 orientations. The four tile types in standard orientation are shown in Figure 4.10, and the edge identifications have already been made. The vertices satisfy $ra = b$, $rb = a$, $rc = d$, $rd = c$, while the edges and tiles have only the condition $r^{10} = 1$. The vertices therefore appear only in the $r = \pm 1$ representations, while the edges and faces appear in all representations. Our boundary and coboundary maps

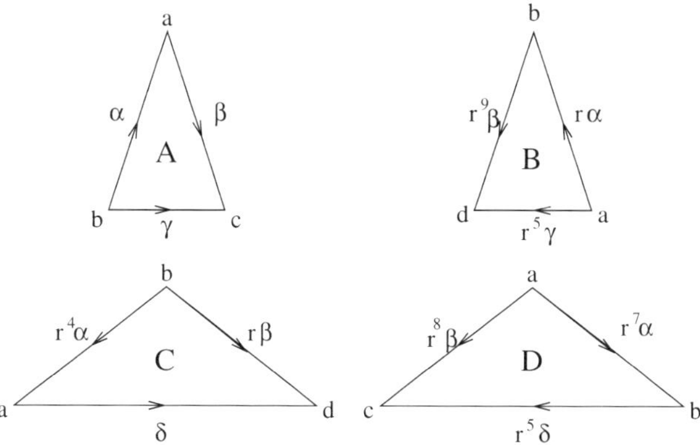

FIGURE 4.10. The four labeled Penrose tiles. Note that $ra = b$, $rb = a$, $rc = d$ and $rd = c$.

are

$$\partial_2 = \begin{pmatrix} -1 & r & r^4 & -r^7 \\ -1 & r^9 & -r & r^8 \\ 1 & -r^5 & 0 & 0 \\ 0 & 0 & 1 & -r^5 \end{pmatrix}, \tag{4.14}$$

$$\delta_1 = \partial_2^T = \begin{pmatrix} -1 & -1 & 1 & 0 \\ r & r^9 & -r^5 & 0 \\ r^4 & -r & 0 & 1 \\ -r^7 & r^8 & 0 & -r^5 \end{pmatrix} \tag{4.15}$$

in all representations, and

$$\partial_1 = \begin{pmatrix} 1-r & -1 & -r & -1 \\ 0 & 1 & 1 & r \end{pmatrix}, \qquad \delta_0 = \partial_1^T = \begin{pmatrix} 1-r & 0 \\ -1 & 1 \\ -r & 1 \\ -1 & r \end{pmatrix} \tag{4.16}$$

in the $r = \pm 1$ representations, and $\delta_0 = 0$ in the other representations (since C^0 is empty in the other representations).

There are four irreducible representations of $\mathbb{Z}_{10}$ over the integers. The scalar ($r = 1$), and pseudoscalar ($r = -1$) representations are 1-dimensional, and there are two 4-dimensional representations. The vector representation has r acting by multiplication on $\mathbb{Z}[r]/(r^4 - r^3 + r^2 - r + 1)$, while the pseudovector representation has r acting by multiplication on the ring $\mathbb{Z}[r]/(r^4 + r^3 + r^2 + r + 1)$.

Scalar representation $r = 1$. In the scalar representation, $C^0 = \mathbb{Z}^2$ (with generators the duals to a and c), $C^1 = \mathbb{Z}^4$, and $C^2 = \mathbb{Z}^4$. Our maps are

$$\delta_0 = \begin{pmatrix} 0 & 0 \\ -1 & 1 \\ -1 & 1 \\ -1 & 1 \end{pmatrix}, \qquad \delta_1 = \begin{pmatrix} -1 & -1 & 1 & 0 \\ 1 & 1 & -1 & 0 \\ 1 & -1 & 0 & 1 \\ -1 & 1 & 0 & -1 \end{pmatrix}. \tag{4.17}$$

We then have $H^0 = \ker \delta_0 = \mathbb{Z}$, with generator $(1,1)^T$, $H^1 = \ker \delta_1 / \operatorname{Im} \delta_0 = \mathbb{Z}$ and $H^2 = \mathbb{Z}^4 / \operatorname{Im} \delta_1 = \mathbb{Z}^2$.

Pseudoscalar representation $r = -1$. In the pseudoscalar representation, we also have $C^0 = \mathbb{Z}^2$, $C^1 = \mathbb{Z}^4$, and $C^2 = \mathbb{Z}^4$, only now our maps are

$$\delta_0 = \begin{pmatrix} 2 & 0 \\ -1 & 1 \\ 1 & 1 \\ -1 & -1 \end{pmatrix}, \qquad \delta_1 = \begin{pmatrix} -1 & -1 & 1 & 0 \\ -1 & -1 & 1 & 0 \\ 1 & 1 & 0 & 1 \\ 1 & 1 & 0 & 1 \end{pmatrix}. \tag{4.18}$$

Both coboundary maps have rank 2, and $\operatorname{Im} \delta_0 = \ker \delta_1$, so $H^0 = H^1 = 0$ and $H^2 = \operatorname{Coker}(\delta_1) = \mathbb{Z}^2$.

Vector and pseudovector representations. In the vector representation, C^0 is trivial while C^1 and C^2 are free modules of dimension 4 over the ring $\mathbb{Z}[r]/(r^4 - r^3 + r^2 - r + 1)$ (or dimension 16 over the integers). The matrix δ_1 has rank 3 over the ring, so H^1 and H^2 each have one copy of the ring. Viewed as abelian groups, each is $\mathbb{Z}^4$.

Finally, in the pseudovector representation, C^0 is trivial while C^1 and C^2 are free of dimension 4 over $\mathbb{Z}[r]/(r^4 + r^3 + r^2 + r + 1)$. The map δ_1 is an isomorphism, so $H^0 = H^1 = H^2 = 0$.

Summary. So far we have only computed the cohomology of the complex Γ_0. However, the substitution map is invertible on chains, hence is invertible on cochains and on cohomology. Thus the direct limit of $H^*(\Gamma_0)$ is simply $H^*(\Gamma_0)$. Adding up the contributions of the different representations, we have the following picture.

(1) $\check{H}^0(\Omega_1) = \mathbb{Z}$, coming from the scalar representation.
(2) $\check{H}^1(\Omega_1)$ contains one copy of the scalar representation and one copy of the (4-dimensional) vector representation. As an abelian group this equals $\mathbb{Z}^5$, but just saying $\check{H}^1 = \mathbb{Z}^5$ misses out on the rotational structure.
(3) $\check{H}^2(\Omega_1)$ has two copies of the scalar representation, 2 copies of the pseudoscalar representation, and one copy of the vector representation. As an abelian group, this means $\mathbb{Z}^8$.
(4) The cohomology of Ω_0 is the rotationally invariant part of the cohomology of Ω_1, so $\check{H}^0(\Omega_0) = \mathbb{Z}$, $\check{H}^1(\Omega_0) = \mathbb{Z}$ and $\check{H}^2(\Omega_0) = \mathbb{Z}^2$.
(5) The rational cohomology of Ω_{rot} is that of $\Omega_0 \times S^1$, but there are two exceptional fibers of the fibration $\Omega_{rot} \to \Omega_0$, each with length $2\pi/5$. These corresponding to the two tilings in Figure 4.11, and give rise

to a $\mathbb{Z}_5$ factor in $\check{H}^2(\Omega_{rot})$. That is, $\check{H}^0(\Omega_{rot}) = \mathbb{Z}$, $\check{H}^1(\Omega_{rot}) = \mathbb{Z}^2$, $\check{H}^2(\Omega_{rot}) = \mathbb{Z}^3 \oplus \mathbb{Z}_5$, $\check{H}^3(\Omega_{rot}) = \mathbb{Z}^2$.

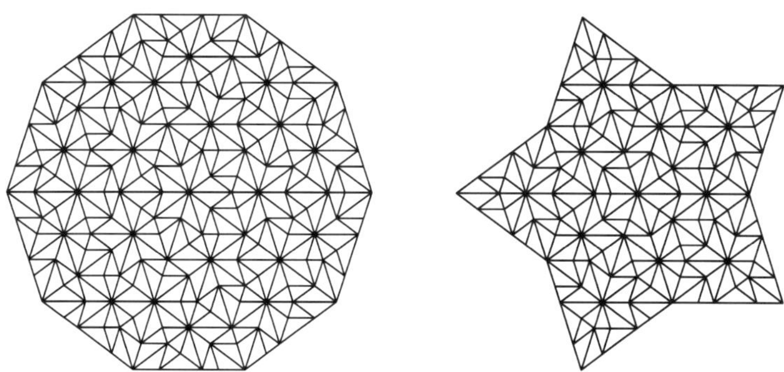

FIGURE 4.11. Central patches of two Penrose tilings with 5-fold rotational symmetry, each of which substitutes into a rotated version of the other

CHAPTER 5

Pattern-equivariant cohomology

By now you are probably convinced that the Čech cohomology of a tiling space is the best thing since sliced bread. Or at least you *should* feel that way. However, you probably don't have a very good feel for what Čech cohomology actually is. If we compute that $\check{H}^2$ of a tiling space is $\mathbb{Z}$, how can we visualize a generator of this group? It's not represented by a differential form. It's not represented by a cellular cochain that assigns integer values to faces. What in blazes *is* it?

Kellendonk and Putnam [**Kel2, KP**] devised a remarkable cohomology theory, called *pattern-equivariant* (or PE) cohomology. To each tiling T, viewed as a marked copy of the plane (or more generally of $\mathbb{R}^n$), they associated a set of differential forms. The closed forms modulo the exact forms define the pattern-equivariant cohomology. They then proved that the PE cohomology of the tiling T is isomorphic to the Čech cohomology of the tiling *space* Ω_T with values in $\mathbb{R}$. In other words, every class in $\check{H}^*(\Omega_T, \mathbb{R})$ can be represented by a differential form on the plane that somehow reflects the structure of T.

This is fantastic! We can finally get a feel for what cohomology is actually telling us. In fact, I cheated in writing the last chapter; the descriptions of one class counting points, another class counting supertiles, and so on, already assumed this identification. There are just two problems:

(1) Working with forms, and hence with real coefficients, we lose all information about divisibility and torsion. The groups $\mathbb{Z}[1/2]$ and $\mathbb{Z}$ look the same when tensored with $\mathbb{R}$, and $\mathbb{Z}_n \otimes \mathbb{R} = 0$.
(2) Although the statement of Kellendonk and Putnam's theorem is simple and intuitive, the proof is not. It relies on some fairly powerful machinery about foliated spaces.

In this chapter we will lay out the formalism of pattern-equivariant cohomology for simple tiling spaces. We will speak of tilings of the plane, but the same constructions work in arbitrary dimensions. We will then give a heuristic argument for the Kellendonk-Putnam theorem. This argument overlooks some important technical details, but correctly describes why the theorem is true. We then consider a variant of PE cohomology with integer coefficients. Remarkably, the technical difficulties in proving the real-valued Kellendonk-Putnam theorem go away, and we are left with a rigorous yet elementary proof that the PE cohomology of T with integer coefficients is isomorphic to the Čech cohomology of Ω_T.

61

Once we understand PE cohomology of simple tiling spaces, we can generalize. We can consider PE cohomology for tiling spaces with rotational symmetry, be it the discrete symmetry of the chair tiling space or the continuous symmetry of the pinwheel tiling space. There are two principal approaches to this last question, one by Kellendonk and one by Rand, and we will go over both.

5.1. Pattern-equivariant functions

Consider a tiling T of the plane, viewed as a marked copy of $\mathbb{R}^2$. A smooth function $f : \mathbb{R}^2 \to \mathbb{R}$ is *pattern-equivariant with radius R* if, for any two points x, x' with $[B_R(x) - x] = [B_R(x') - x']$, we have $f(x) = f(x')$. In other words, the value at x of a PE function of radius R is determined by the pattern of the tiling on $B_R(x)$.

Let $C^\infty_{P,R}(T)$ be the set of PE functions of radius R on T. Let $C^\infty_P(T) = \cup_R C^\infty_{P,R}(T)$. Functions in $C^\infty_P(T)$ are called *strongly PE*. The radius R can be arbitrarily large, but must be finite. This definition is analogous to that of a function of compact support; the support can be arbitrarily large, but cannot extend to infinity.

Let $C^\infty_{P,w}(T)$ be the set of functions that are uniform limits of strongly PE functions, and whose derivatives of all orders are uniform limits of strongly PE functions. These functions are called *weakly PE*. This is analogous to functions that may not be compactly supported, but whose values and derivatives decay at infinity.

A (strongly or weakly) PE 1-form is an expression of the form $f_1(x)dx^1 + f_2(x)dx^2$, where f_1 and f_2 are (strongly or weakly) PE. A PE 2-form takes the form $f_{12}(x)dx^1 dx^2$ with $f_{12}(x)$ PE. We denote the set of strongly (resp. weakly) k-forms by $\Lambda^k_P(T)$ (resp. $\Lambda^k_{P,w}(T)$).

It is easy to check that the exterior derivative of a PE form is PE. Since $d^2 = 0$ on all forms, we have differential complexes

$$0 \to \Lambda^0_P(T) \xrightarrow{d} \Lambda^1_P(T) \xrightarrow{d} \Lambda^2_P(T) \to 0 \tag{5.1}$$

and

$$0 \to \Lambda^0_{P,w}(T) \xrightarrow{d} \Lambda^1_{P,w}(T) \xrightarrow{d} \Lambda^2_{P,w}(T) \to 0. \tag{5.2}$$

The cohomology of the first complex is the strongly PE cohomology of T, denoted $H^*_P(T)$, and the cohomology of the second complex is the weakly PE cohomology of T, denoted $H^*_{P,w}(T)$. In addition to the strong and weak PE cohomologies, we can consider the *mixed* PE cohomology, meaning the closed strongly PE forms modulo the strongly PE forms that are d of weakly PE forms.

EXERCISE 5.1. *Prove that the exterior derivative of a strongly PE form is strongly PE.*

EXERCISE 5.2. *Prove that the exterior derivative of a weakly PE form is weakly PE.*

EXERCISE 5.3. *Let T be the 1-dimensional "one black tile" tiling. Characterize the strongly and weakly PE functions on T. Find a smooth function on T that is the uniform limit of strongly PE functions, but whose derivative is not weakly PE.*

Note that every closed form on $\mathbb{R}^2$ (and hence every closed PE form) is exact. However, it may not be d of a PE form! The forms dx^1 and dx^2 are PE, and are the derivatives of the functions x^1 and x^2, respectively, but x^1 and x^2 are not PE, so dx^1 and dx^2 always represent nonzero classes in $H^1_P(T)$. These forms also represent nontrivial classes in $H^1_{P,w}(T)$. Likewise $dx^1 dx^2$ represents a nontrivial class in $H^2_P(T)$.

Weak PE cohomology is not yet well understood. However, the key properties of strong and mixed PE cohomology are described in the following two theorems. The first is due to Kellendonk and Putnam [**KP**]:

THEOREM 5.1. *If T is a simple tiling, then $H^k_P(T)$ is isomorphic to $\check{H}^k(\Omega_T, \mathbb{R})$.*

COROLLARY 5.2. *If Ω_T is minimal and if $T' \in \Omega_T$, then $H^k_P(T') = H^k_P(T)$.*

The second theorem, due to Kellendonk [**Kel2**], applies PE cohomology to the tiling deformations of chapter 3:

THEOREM 5.3. *The asymptotically negligible classes of a tiling space are precisely those that are represented by closed strongly PE forms that are d of weakly PE forms. In other words, the infinitesimal deformations of a tiling, up to topological conjugacy, are described by the first mixed PE cohomology of the tiling with values in $\mathbb{R}^d$.*

5.2. How to view pattern equivariance

Recall that a point on the n-th Gähler approximant Γ_n to Ω_T describes how to lay n layers of tiles around the origin. If $n > 0$, there exist radii r_n and R_n such that those layers are always contained in $B_{R_n}(0)$ and always contain $B_{r_n}(0)$, and r_n and R_n go to infinity as $n \to \infty$. Also recall that $\pi_n : \Omega_T \to \Gamma_n$ is the natural projection that takes a tiling and describes its first n layers around the origin. Restricting to the orbit of T, we get a map (also denoted π_n) from $\mathbb{R}^2$ to Γ_n. This map is onto, since every pattern in every tiling in Ω_T is found somewhere in T.

EXERCISE 5.4. *Prove the above assertion that r_n and R_n exist and go to infinity as $n \to \infty$.*

THEOREM 5.4. *If $f : \Gamma_n \to \mathbb{R}$ is a smooth function, then $\pi_n^* f : \mathbb{R}^2 \to \mathbb{R}$ is strongly PE with radius R_n. Conversely, if $g : \mathbb{R}^2 \to \mathbb{R}$ is strongly PE with radius $R < r_n$, then $g = \pi_n^* f$ for some $f : \Gamma_n \to \mathbb{R}$. The set of all strongly PE functions on T equals the union of the sets of pullbacks of smooth functions on all approximants Γ_n.*

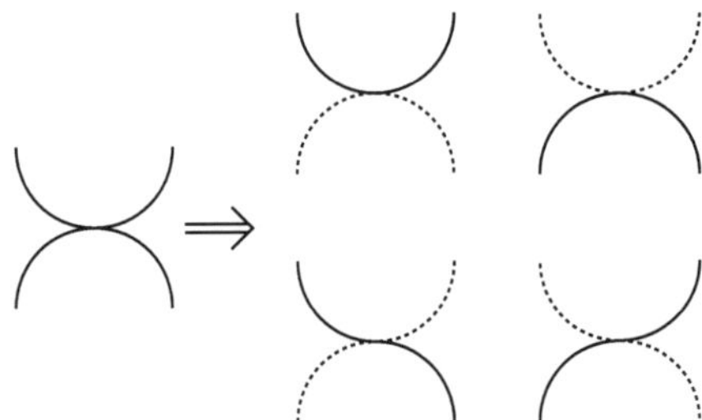

FIGURE 5.1. The neighborhood of a branch point (left) is described by several overlapping charts (right).

PROOF. If f is a function on Γ_n and if $[B_{R_n}(x)] - x = [B_{R_n}(y)] - y$, then the n layers of tiles around x and y are the same, and $\pi_n(x) = \pi_n(y)$, so $\pi_n^* f = f \circ \pi_n$ takes on the same values at x and y. Now suppose that g is a PE function with radius less than r_n (and hence is PE with radius r_n). If $\pi_n(x) = \pi_n(y)$, then $[B_{r_n}(x)] - x = [B_{r_n}(y)] - y$, so $g(x) = g(y)$. We then define $f(\pi_n(x)) = g(x)$. Since π_n is onto, this defines a function on all of Γ_n whose pullback is g. Since every strongly PE function is the pullback of a function on Γ_n for all sufficiently large n, and since all pullbacks are strongly PE, the set of strongly PE functions equals the set of pullbacks. $\qquad\square$

The only subtlety to this argument is the concept of a smooth function on Γ_n. Smoothness is usually defined only for smooth manifolds, and Γ_n is a *branched* manifold. Near a branch, a branched manifold is described by several charts, as in Figure 5.1. A function is considered smooth if it is smooth with respect to all of these charts. In other words, the value of the function and all of its derivatives on the branch set must be the same for every choice of which way to go. It is not hard to check that a function on Γ_n is smooth if and only if its pullback to $\mathbb{R}^2$ is smooth.

Likewise, we can define differential forms on Γ_n and consider the de Rham cohomology of Γ_n. For manifolds, the de Rham theorem says that the de Rham cohomology is isomorphic to the ordinary (simplicial, singular, cellular or Čech) cohomology with real coefficients. The analogous theorem applies to branched manifolds (see [**Sa3**]), but this theorem had not appeared in the literature when PE cohomology was defined. We will skip the technical details needed to prove this generalized de Rham theorem.

PROOF OF THEOREM 5.1. Once we identify PE forms with forms on Γ_n and accept the generalized de Rham theorem, Theorem 5.1 is almost a tautology.

$$
\begin{aligned}
H_P^*(T) &= \frac{\text{Closed PE forms on } T}{d(\text{PE forms on } T)} \\
&= \frac{\varinjlim \text{Closed forms on } \Gamma_n}{\varinjlim \text{Exact forms on } \Gamma_n} \\
&= \varinjlim H_{deRham}^*(\Gamma_n)
\end{aligned}
$$

$$\begin{aligned} &= \varinjlim \check{H}^*(\Gamma_n, \mathbb{R}) \\ &= \check{H}^*(\Omega_T, \mathbb{R}). \end{aligned} \tag{5.3}$$

$\square$

5.3. Integer coefficients

To make the switch to integer coefficients [**Sa3**], we use the fact that T isn't just a pattern. It is a *tiling*, and a tiling by polygons at that. T defines a CW decomposition of $\mathbb{R}^2$ into vertices, edges and faces. A k-cochain on T is just a function that assigns an integer to every k-cell of T.

The definition of pattern equivariance is almost as before. To each k-cell c, let v_c be the center-of-mass of c. We say a k-cochain α is PE with radius R if, whenever two k-cells c and c' have $[B_R(v_c)] - v_c = [B_R(v_{c'})] - v_{c'}$, then $\alpha(c) = \alpha(c')$. In other words, the value of α on a k-cell can only depend on the pattern of radius R around the center of that cell.

EXERCISE 5.5. *Suppose that all tiles in T have diameter D or less, and that the k-cochain α is PE with radius R. Show that the coboundary of α is PE with radius $R + D$.*

A cochain is PE if it is PE for some radius R. There is no distinction between strongly PE and weakly PE cochains, since the only way a sequence of cochains can converge uniformly is if it is eventually constant. A PE cochain is called closed if its coboundary is zero, and exact if it is the coboundary of a PE cochain. The integer-valued PE cohomology, denoted $H_P^*(T, \mathbb{Z})$, is the quotient of the closed PE cochains by the exact PE cochains.

The approximants Γ_n are also CW complexes, with a CW structure defined by the various tiles, edges and vertices. The map π_n from $\mathbb{R}^2$ to Γ_n is cellular, and we can consider pullbacks of cellular cochains on Γ_n.

EXERCISE 5.6. *By modifying the proof of Theorem 5.4, show that a cochain on T is PE if and only if it is the pullback of a cochain on Γ_n for some n.*

The interpretation of PE cohomology in terms of inverse limits is exactly as before. Instead of being the direct limit of the de Rham cohomology of the approximants Γ_n, it is the direct limit of the cellular cohomology of Γ_n. However, the cellular cohomology of a CW complex is isomorphic to the Čech cohomology of Γ_n, and the direct limit of the Čech cohomology of Γ_n is the Čech cohomology of Ω_T. The fact that Γ_n is not a manifold does not complicate things at all!

Instead of using integer coefficients, we can take our coefficients to lie in any discrete Abelian group, and the argument is identical. In summary, we have proven:

THEOREM 5.5. *Let G be a discrete Abelian group. The PE cohomology of a simple tiling T with coefficients in G is isomorphic to the Čech cohomology of Ω_T with coefficients in G.*

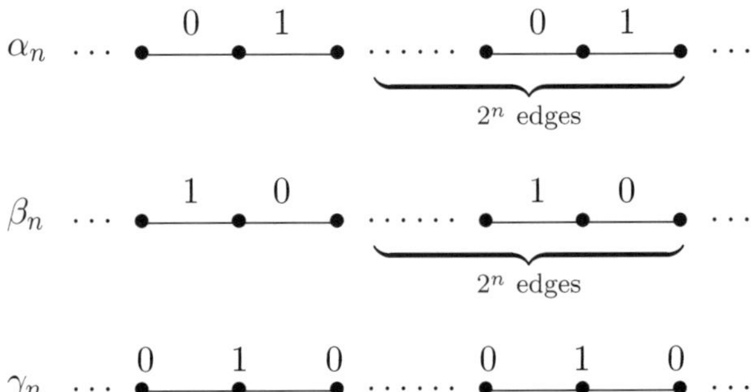

FIGURE 5.2. The 1-cochains α_n and β_n, and the 0-cochain γ_n whose coboundary is $\beta_n - \alpha_n$

5.4. Interpreting tiling cohomology

We return to the arrow tiling and find explicit representatives of the various cohomology classes. In all cases we are looking at the cohomology of $\Omega_T = \Omega_1$, not Ω_0 or Ω_{rot}.

First we need the CW structure of an arrow tiling T. Geometrically, T is an infinite checkerboard with extra markings. We orient all the horizontal edges from left to right, orient the vertical edges from bottom to top, and give each face the standard orientation where $dx^1 dx^2$ is a positive form.

There isn't much to do in dimension zero. $\check{H}^0(\Omega_T) = \mathbb{Z}$, and the generator is the function that equals 1 on every vertex.

$\check{H}^1(\Omega_T) = \mathbb{Z}[1/2]^2$. If a and b are integers, then the class (a, b) is represented by a cochain that evaluates to a on every horizontal edge and b on every vertical edge. We need more than that, however. We need to represent $(a/2^n, b/2^m)$.

Each horizontal edge of an order-n supertile consists of 2^n ordinary edges. Pick *one* of these ordinary edges – say the rightmost one, and use the same prescription for all horizontal edges of all order-n supertiles. Consider the cochain α_n that evaluates to 1 on this special edge and to zero on all of the other $2^n - 1$ edges (and to zero on all vertical edges). See Figure 5.2.

Let β_n be defined similarly, only with a different choice of special edge (say, the next-to-rightmost). The cochain $\beta_n - \alpha_n$ is exact, being the coboundary of a 0-cochain γ_n that evaluates to 1 on the common vertex of the two special edges and zero on all other vertices. In other words, the cohomology class of α_n does not depend on the choice of special edge.

Now look at $2^n \alpha_n$. This is cohomologous to the sum of all of the 2^n different choices of special edge. But that is the cochain that evaluates to 1 on *all* horizontal edges, and corresponds to $(1, 0)$. In other words, α_n represents $(1/2^n, 0)$. Representing $(0, 1/2^m)$ is similar. The class $(a/2^n, b/2^m)$ is represented by a cochain that evaluates to a at every 2^n-th horizontal edge

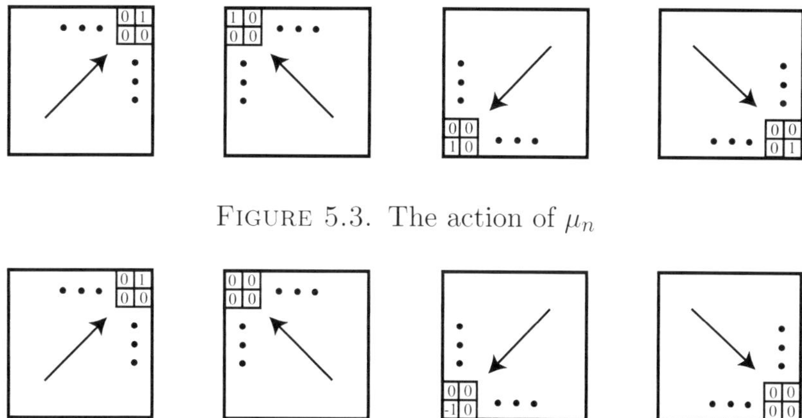

FIGURE 5.3. The action of μ_n

FIGURE 5.4. The action of $\nu_{1,n}$

(as defined by the supertile structure), evaluates to b at every 2^m-th vertical edge, and that evaluates to zero everywhere else.

The situation for $\check{H}^2$ is similar. Pick a special tile from each order-n supertile, say at the head of the arrow. Since $\check{H}^2(\Omega_T) = \frac{1}{3}\mathbb{Z}[1/4] \oplus \mathbb{Z}[1/2]^2$, we must find representatives for $(4^{-n}/3, 0, 0)$, $(0, 2^{-n}, 0)$ and $(0, 0, 2^{-n})$.

Let μ_n evaluate to 1 at the special tile in each order-n arrow supertile, as in Figure 5.3. Note that μ_0, which evaluates to 1 on every tile, represents $(1, 0, 0)$. There are four order-n supertiles in each order-$(n+1)$ supertile, so μ_n is cohomologous to $4\mu_{n+1}$. Since μ_0 is cohomologous to $4^n \mu_n$, μ_n represents $(4^{-n}, 0, 0)$. To get a representative for $(4^{-n}/3, 0, 0)$, we note that each order-n chair supertile consists of three order-n arrow supertiles, and that these can be distinguished by collaring out to a finite distance. Let μ'_n evaluate to 1 on the special tile in one of these three supertiles, and to zero everywhere else. $3\mu'_n$ is cohomologous to μ_n, so μ'_n represents $(4^{-n}/3, 0, 0)$.

Let $\nu_{1,n}$ evaluate to 1 on the special tile in each order-n supertile whose arrow points northeast, -1 on the special tile in supertiles whose arrows points southwest, and 0 on all other tiles, as in Figure 5.4. Let $\nu_{2,n}$ evaluate to 1 on the special tile in supertiles whose arrow points northwest, -1 on the special tile in supertiles whose arrows points southeast, and 0 on all other tiles, as in Figure 5.5. Since each order-$(n+1)$ supertile with arrow pointing in a direction contains two order-n supertiles with arrows pointing in the same direction, plus two more whose arrows are antiparallel to each other, $\nu_{1,n}$ is cohomologous to $2\nu_{1,n+1}$ and $\nu_{2,n}$ is cohomologous to $2\nu_{2,n+1}$. $\nu_{1,n}$ represents $(0, 2^{-n}, 0)$ and $\nu_{2,n}$ represents $(0, 0, 2^{-n})$.

5.5. Gap labeling and K-theory

Tilings have been studied with the machinery of algebraic K-theory. From the tiling space Ω_T and the action of the translation group, one defines a C^* algebra of functions. The K-theory of this algebra is related to

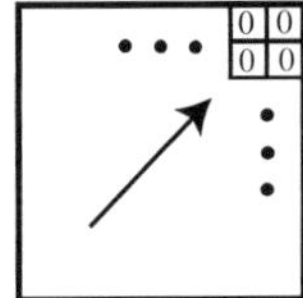 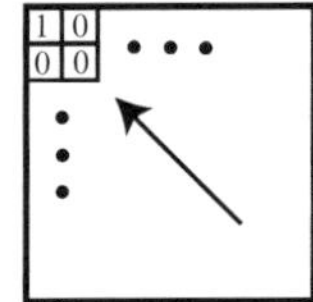 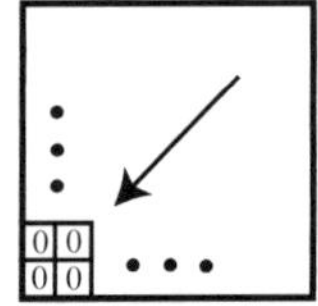 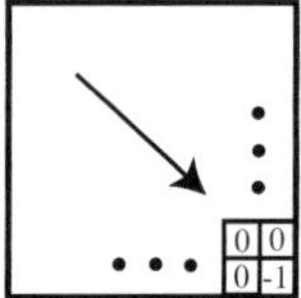

FIGURE 5.5. The action of $\nu_{2,n}$

the integer-valued Čech cohomology of Ω_T. K_0 is the direct sum of the cohomologies in even codimensions (i.e., $\check{H}^0 \oplus \check{H}^2$ when dealing with tilings of the plane, and $\check{H}^1$ when dealing with tilings of the line), while K_1 is the direct sum of the cohomologies in odd codimension. There is a trace operation that maps K_0 to $\mathbb{R}$. The image of this map is a countable additive subgroup of $\mathbb{R}$ and is called the *gap-labeling group* of Ω_T.

The gap-labeling group can help distinguish tiling spaces. Even if the spaces have identical cohomologies (and hence identical K_0), their trace operations and gap-labeling groups may be different. For instance, a space with $K_0 = \mathbb{Z}^2$ might have gap-labeling group $\mathbb{Z} \oplus \sqrt{2}\mathbb{Z}$ or $\mathbb{Z} \oplus \sqrt{3}\mathbb{Z}$. These are isomorphic as abstract groups, but differ as subgroups of $\mathbb{R}$.

The gap-labeling group also has direct physical significance, thanks to the following theorem of Bellissard, Benedetti and Gambaudo [**BBG**].

THEOREM 5.6 (Gap labeling). *Let T be a simple tiling of the plane with well-defined patch frequencies, and let V be a strongly PE function on T. Consider the Hamiltonian operator $H = -\frac{\nabla^2}{2} + V$. The spectrum of H has gaps. For $E \in \mathbb{R}$, let $\rho(E)$ be the number of states, per unit area, with energy less than E. (This is called the integrated density of states.) If E is not in the spectrum of H, then $\rho(E)$ lies in the gap-labeling group of Ω_T.*

In other words, the possible integrated densities of states at the gaps in the spectrum (which can be used to label the gaps) can be computed from T without any further knowledge of the potential V. If the tiling represents the positions of atoms in a quasicrystal, then the electrons are governed by a Hamiltonian operator as described above, and the K-theory of Ω_T gives important information about the electrical properties of the quasicrystal!

The subtlety is in describing the trace map and computing the gap-labeling group. This is most easily done using PE cohomology.

5.6. Averages and Ruelle-Sullivan currents

To each PE function we can associate a number – the average of the function over the entire plane. As long as the tiling has well-defined patch frequencies, this average is well-defined. Likewise, we can take the average of a differential form. If $\alpha = f_1 dx^1 + f_2 dx^2$, we define $\langle \alpha \rangle = \langle f_1 \rangle dx^1 + \langle f_2 \rangle dx^2$, where the angle brackets denote averages. Kellendonk and Putnam showed that the average of a PE-exact form is zero, so taking averages gives a group homomorphism from H_{PE}^* to the exterior algebra of $\mathbb{R}^d$, and hence from

$\check{H}^*(\Omega_T, \mathbb{R})$ to the exterior algebra of $\mathbb{R}^d$. This homomorphism is onto (just consider constant PE forms), and is actually a homomorphism of rings. For details, see [**KP**].

Now restrict this homomorphism to the integer-valued cohomology. The image $\check{H}^k$ has units of $(\text{Length})^{-k}$, so the different dimensions are easily distinguished. The image of the 2-dimensional cohomology is a density, with units of $(\text{Area})^{-1}$. In fact, this image is the gap-labeling group, and the averaging operation is the abstract trace map from K-theory.

(In measurable dynamical systems with an $\mathbb{R}$ action, there is a structure called a Ruelle-Sullivan current that maps H^1 to the real numbers. In the case of $\mathbb{R}^n$ actions by translation on n-dimensional tilings, Kellendonk and Putnam generalized this structure and showed that is was the same as the averaging PE forms over a single tiling.)

We can also describe the gap-labeling group directly using integer-valued cochains. If β is a 2-cochain, we define the average

$$\langle \beta \rangle = \lim_{R \to \infty} \frac{\beta([B_R(0)])}{\text{Area of } [B_R(0)]}. \tag{5.4}$$

The use of balls centered at the origin (as opposed to other shapes) is not important. The limit is the same for any increasing sequence of patches that exhaust the plane and whose perimeter grows slower than the area. In practice, we often use supertiles of increasing order.

EXERCISE 5.7. *Show that two representatives of the same cohomology class have the same average.*

Examples.

(1) In the chair tiling, the top cohomology is generated by μ'_n, $\nu_{1,n}$ and $\nu_{2,n}$. Applied to a supertile of order $m > n$, $\nu_{1,n}$ and $\nu_{2,n}$ evaluate to at most 2^{m-n}. However, the area of the supertile scales like 4^m, so the average values of $\nu_{1,n}$ and $\nu_{2,m}$ are zero. The cochain μ_n manifestly has average 4^{-n}, and $3\mu'_n$ is cohomologous to μ_n, so $\langle \mu'_n \rangle = 4^{-n}/3$, and the gap-labeling group is $\frac{1}{3}\mathbb{Z}[1/4]$.

(2) Another way to see that $\nu_{1,n}$ cannot contribute is from its rotational properties. Rotating the tiling by 180 degrees sends $\nu_{1,n}$ to $-\nu_{1,n}$, but preserves the overall patch frequencies. Thus $\langle \nu_{1,n} \rangle = -\langle \nu_{1,n} \rangle = 0$. This generalizes easily: *If $\check{H}^2$ can be written as the direct sum of irreducible representations of the rotation group, then only the rotationally invariant part of $\check{H}^2$ contributes to the gap-labeling group.*

(3) The Penrose tiling has $\check{H}^2 = \mathbb{Z}^8$, but the rotationally invariant part of $\check{H}^2$ is just $\mathbb{Z}^2$, with generators that count the two different kinds of triangle. Since the narrow triangles outnumber the wide triangles by a factor of $\tau : 1$, the gap-labeling group is $c(\mathbb{Z} \oplus \tau\mathbb{Z})$, where τ is the golden mean and c is a normalization constant that depends on the overall size of the tiles.

(4) Gab-labeling groups of 1-dimensional tiling spaces are similar, only coming from $\check{H}^1$ instead of $\check{H}^2$. The Thue-Morse tiling space has $\check{H}^1 = \mathbb{Z} \oplus \mathbb{Z}[1/2]$. The $\mathbb{Z}$ factor is invariant under substitution. Since it takes the same value on a supertile of size 2^n as on a single tile, it averages to zero. The $\mathbb{Z}[1/2]$ factor maps nontrivially, and the gap-labeling group is $\mathbb{Z}[1/2]/L$, where L is the length of a single tile.

(5) The Fibonacci tiling has $\check{H}^1 = \mathbb{Z}^2$, with the generators counting the two kinds of tiles. Since b tiles outnumber a tiles by $\tau : 1$, the gap-labeling group is $c(\mathbb{Z} \oplus \tau\mathbb{Z})$, like that of the Penrose tiling.

(6) The Fibonacci tiling is a cut-and-project tiling as well as a substitution tiling. Let α be an irrational number and let Ω_α be the space of 1-dimensional cut-and-project tilings obtained from the procedure of section 1.4. The cohomology group $\check{H}^1(\Omega_\alpha) = \mathbb{Z}^2$ does not depend on α. The two kinds of tiles appear in a ratio of $\alpha : 1$ and the gap-labeling group is $\mathbb{Z} \oplus \alpha\mathbb{Z}$, up to an overall scale. The gap-labeling groups of Ω_α and $\Omega_{\alpha'}$ are isomorphic (as subgroups of $\mathbb{R}$) if and only if α and α' are related by a Mobius transformation: $\alpha' = \frac{a\alpha+b}{c\alpha+d}$, with a, b, c, and d integers with $ad - bc = \pm 1$. In that case, Ω_α and $\Omega_{\alpha'}$ are actually topologically conjugate (up to an overall scale). In short, cohomology is useless for distinguishing between the different spaces Ω_α, but the gap-labeling group completely classifies these spaces up to topological conjugacy.

5.7. PE cohomology and rotations

So far we have considered the PE cohomology of *simple* tilings, where each tile appears in on a finite number of orientations. How are we to construct the PE cohomology of the pinwheel tiling?

At first glance this appears impossible. The pinwheel tiling itself is 2-dimensional, but the pinwheel tiling space has 3 continuous degrees of freedom, and has a nontrivial $\check{H}^3$. There is no hope for representing this 3-dimensional class with the 0-, 1-, and 2-forms that live on $\mathbb{R}^2$.

Two approaches have been suggested for overcoming this problem. Kellendonk proposes to consider PE forms on the Euclidean group, rather than on $\mathbb{R}^2$. The resulting cohomology theory is isomorphic to the Čech cohomology of Ω_{rot} with real coefficients. An integer version of Kellendonk's cohomology also exists.

Rand [**Ran**] works with forms on $\mathbb{R}^2$ and uses representations of the rotation group to define pattern-equivariance. Each representation yields a different cohomology theory. None of these theories is isomorphic to the Čech cohomology of Ω_{rot}, but the trivial representation gives a cohomology theory isomorphic to $\check{H}^*(\Omega_0)$.

Kellendonk's approach. As a warmup to Kellendonk's approach, let's revisit pattern-equivariant cohomology for simple tilings. Let T be a tiling, and consider functions on $\mathbb{R}^2$. Here we do not think of $\mathbb{R}^2$ as the space on which T lives, but rather as the group of translations that acts on T. We say a function f on $\mathbb{R}^2$ is pattern-equivariant (with respect to T) with radius R if, whenever $T - x$ and $T - x'$ agree on $B_R(0)$, then $f(x) = f(x')$. This is exactly the same definition as before, only now we think of x and x' as translations rather than as points in T.

This approach generalizes to an arbitrary group that acts on tilings, and in particular to the Euclidean group. If a group G acts on our tilings, we consider smooth functions $f : G \to \mathbb{R}$, and call such a function PE with radius R if, whenever we have elements $g, g' \in G$ such that $g(T)$ and $g'(T)$ agree on $B_R(0)$, then $f(g) = f(g')$. As before, we consider PE functions of arbitrary radius. The algebra of PE forms is then generated by the PE functions and the exterior derivatives of the PE functions. If G is the Euclidean group, then T needn't be a simple tiling. It can be a pinwheel tiling, for example, with tiles pointing in an infinite number of different directions.

The tiling T defines a map $\hat{T} : G \to \Omega_{rot}$, namely $\hat{T}(g) = g(T)$. If Γ^n is an approximant to Ω_{rot} (in either the Gähler or Anderson-Putnam construction) and if π_n is the natural projection from Ω_{rot} to Γ^n, then $\pi_n \circ \hat{T}$ is a map from G to Γ^n.

EXERCISE 5.8. *Show that a function f on G is PE if and only if there exists an integer N such that, for all $n > N$, f is the pullback by $\pi_n \circ \hat{T}$ of a function on Γ^n.*

EXERCISE 5.9. *Show that a form α on G is PE if and only if there exists an integer N such that, for all $n > N$, α is the pullback by $\pi_n \circ \hat{T}$ of a form on Γ^n.*

EXERCISE 5.10. *Show that the pattern-equivariant cohomology (with rotations, in the sense of Kellendonk) of T is isomorphic to the Čech cohomology of Ω_{rot} with real coefficients. You may take the de Rham theorem for branched manifolds as given.*

Defining an integer version of this cohomology is slightly tricky, since we need to induce a CW decomposition of G from the tiling T. The vertices, edges and faces of T naturally define one, two, and three-dimensional regions in G, respectively, but these regions are not contractible. Topologically, they are circles, annuli and solid tori. To get a CW decomposition of G we need to introduce additional vertices, edges and faces that break these regions into contractible cells.

Here is one way to do that. We call $g \in G$ a vertex if the origin is a vertex of $g(T)$ and $g(T)$ has a horizontal edge pointing out from the origin (i.e., an edge parallel to the x^1 axis). We say $g \in G$ is on an edge if the origin is a vertex of $g(T)$, or if the origin sits on a horizontal edge of $g(T)$. We say $g \in G$ is on a face if the origin is on an edge of $g(T)$, or is in the interior of a tile with a horizontal edge.

Once we have a cellular decomposition of G defined in terms of the pattern of $g(T)$ near the origin, we can define pattern-equivariant cochains to be those that depend only on a finite region of $g(T)$, where g is an arbitrary point in the cell being evaluated. It is not hard to check that the coboundary of a PE cochain is PE, and that the resulting PE cohomology of G is well-defined.

EXERCISE 5.11. *Modify the proof the Theorem 5.5 to show that the PE cohomology of G with integer coefficients is isomorphic to the Čech cohomology of Ω_{rot}.*

Rand's approach. Let G_E be the Euclidean group, let G be a closed subgroup of G_E that contains all translations, and let $G_0 \subset G$ be the rotational subgroup of G, namely those elements that send the origin to itself. G_0 might be trivial, if G consists entirely of translations, or G_0 might be finite (e.g. if we look at all translations and rotations by multiples of 90 degrees acting on the chair tiling), or G_0 might be $SO(2)$. Rand's cohomology requires an explicit choice of G, a vector space V, and a representation $\rho : G_0 \to GL(V)$. Since G_0 is compact, we can always choose an inner product on V such that ρ takes values in $O(V)$.

A function $f : \mathbb{R}^2 \to V$ is called PE with representation ρ (denoted PE_ρ) with radius R if, whenever $x, x' \in \mathbb{R}^2$ and $g_0 \in G_0$ satisfy

$$[B_R(0)]_{T-x} = [B_R(0)]_{g_0(T-x')}, \tag{5.5}$$

then

$$f(x) = \rho(g_0)f(x'). \tag{5.6}$$

In other words, whenever $T - x$ and $T - x'$ agree on a ball of radius R around the origin up to a rotation $g_0 \in G_0$, then $f(x)$ and $f(x')$ agree up to $\rho(g_0)$. As usual, we consider functions that are PE_ρ with arbitrarily large radius.

A PE_ρ 2-form is just a PE_ρ function times $dx^1 \wedge dx^2$. However, since the 1-forms dx^1 and dx^2 rotate nontrivially, a PE_ρ 1-form is not just a PE_ρ function times dx^1 or dx^2. If $\mathbf{v} \in \mathbb{R}^2$ is a vector and α is a PE_ρ 1-form, then we require $\alpha(x)(g_0\mathbf{v}) = \rho(g_0)(\alpha(x')(\mathbf{v}))$. Feeding the vector $\mathbf{v}$ to $\alpha(x')$ and then rotating via the representation ρ is not equivalent to feeding $\mathbf{v}$ to $\alpha(x)$. It is equivalent to feeding $g_0(\mathbf{v})$ to $\alpha(x)$.

FIGURE 5.6. Two patches of T related by rotation

This is illustrated in Figure 5.6. The patch of T around x is the same as the patch of T around x', only rotated by 90 degrees. Let g_0 be a counterclockwise rotation by 90 degrees. The vector $\mathbf{v}$ at x' corresponds to the vector $\mathbf{w} = g_0 \mathbf{v}$ at x, so feeding $\mathbf{v}$ to $\alpha(x')$ corresponds (via ρ) to feeding $\mathbf{w}$ to $\alpha(x)$. In particular, if $\alpha(x') = \alpha_1 dx^1 + \alpha_2 dx^2$, then $\alpha(x) = -\rho(g_0)(\alpha_2)dx^1 + \rho(g_0)(\alpha_1)dx^2$.

EXERCISE 5.12. *Show that the exterior derivative of a PE_ρ form is PE_ρ.*

As a result of this exercise, we can define the PE cohomology of T with representation ρ (aka PE_ρ cohomology):

$$H_\rho^*(T, \mathbb{R}) = \frac{\text{Closed PE}_\rho \text{ forms}}{d(\text{PE}_\rho \text{ forms})}. \tag{5.7}$$

One can also define PE_ρ cohomology with integer (or group) coefficients. Instead of taking a vector space V, we need an Abelian group V on which $\rho(g)$ acts for each $g \in G_0$. If $V = \mathbb{Z}^n$, this means that the representation ρ takes values in $GL(n, \mathbb{Z})$ rather than $O(n)$. As before, let v_c be the center-of-mass of the k-cell c, and we consider the possibility that $T - v_c$ and $T - v_{c'}$ agree on $B_R(0)$, up to rotation. If α is a PE_ρ cochain with radius R, and if $[B_R(v_c)] - v_c = g_0([B_R(v_{c'})] - v_{c'})$, then we require $\alpha(c) = \rho(g_0)\alpha(c')$. We then have

$$H_\rho^*(T, \mathbb{Z}) = \frac{\text{Closed PE}_\rho \text{ cochains}}{\delta(\text{PE}_\rho \text{ cochains})}. \tag{5.8}$$

EXERCISE 5.13. *Suppose that ω is a PE_ρ k-form. Show that $\alpha(c) = \int_c \omega$ is a PE_ρ k-cochain. That is, the two notions of PE_ρ are consistent.*

PE_ρ cohomology is certainly well-defined for any representation ρ, but what does it mean? Kellendonk's PE cohomology is isomorphic to Čech cohomology, which is known to be a homeomorphism invariant. However, the functorial properties of PE_ρ are unknown. It is certainly not a homeomorphism invariant, since homeomorphisms do not need to respect the rotational properties of tilings. Is it an invariant of topological conjugacies, or more generally of maps that commute with rotations and send translational orbits to translational orbits? Probably, but this has not been proved.

Indeed, the signal virtue of PE_ρ cohomology is that it is (generally) *not* isomorphic to any other known topological invariant. It is new, and has the potential to give us information about a tiling that is not accessible by other means.

Even so, there are a number of cases where PE_ρ does boil down to known quantities. Here are three important results. Complete proofs may be found in [**Ran**].

THEOREM 5.7. *If ρ is the trivial 1-dimensional representation, then $H_\rho^*(T, \mathbb{R})$ is isomorphic to $\check{H}_\rho^*(\Omega_0, \mathbb{R})$.*

(The same theorem also holds with integer coefficients, although the proof in [**Ran**] is restricted to forms.)

SKETCH OF PROOF. If 1 denotes the trivial representation, then PE_1 functions on T are simply pullbacks of functions on approximants to Ω_0. Likewise, PE_1 forms on T are pullbacks of forms on approximants to Ω_0. $\square$

Recall that, for simple tilings, $\check{H}^*(\Omega_0, \mathbb{R})$ is the rotationally invariant part of $\check{H}^*(\Omega_1, \mathbb{R})$. In those cases, Theorem 5.7 can be restated as saying that the rotationally invariant part of $\check{H}^*(\Omega_1, \mathbb{R})$ corresponds to the PE cohomology with the trivial representation.

The rest of $\check{H}^*(\Omega_1, \mathbb{R})$ is recovered by working with other representations:

THEOREM 5.8. *If T is a simple tiling, then $\check{H}^*(\Omega_1, \mathbb{R})$ is the direct sum of terms corresponding to irreducible representations of G_0. The term corresponding to the irreducible representation ρ is precisely $H_\rho^*(T, \mathbb{R})$.*

Again, there is a version of this theorem with integer coefficients, but $\check{H}^*(\Omega_1, \mathbb{Z})$ does not necessarily decompose as the direct sum of pieces coming from different irreducible representations. However, we don't have to work with irreducible representations. In particular, if G_0 is a finite group, then we can consider the highly reducible *left regular representation* whose dimension is the cardinality of G_0. If e_g is the basis element corresponding to g, then $\rho(h)e_g = e_{hg}$. Over the reals, the left-regular representation is the direct sum of every irreducible representation counted once. Over the integers, things are more subtle, and the left regular representation is not generally a direct sum.

THEOREM 5.9. *If T is a simple tiling and ρ is the left regular representation of G_0, then $H_\rho^*(T, \mathbb{Z})$ is isomorphic to $\check{H}^*(\Omega_1, \mathbb{Z})$.*

CHAPTER 6

Tricks of the trade

In previous chapters we laid out a general theory on how to compute the cohomology of a substitution tiling space, and what to do with that information. However, that's not how people *really* compute things! At least not usually. There are lots of tricks for making our computations easier, as we'll see in this chapter.

6.1. One dimensional methods

Life in one dimension is simplified by the lack of geometry. All tiles are intervals (albeit sometimes of different length) and the substitution is described completely by combinatorial data. In many cases, computing $\check{H}^1$ of a tiling space reduces to just studying the substitution matrix.

Proper substitutions. A substitution σ in one dimension is called *proper* if every substituted letter $\sigma(a_i)$ begins with the same letter, and if every substituted letter ends with the same letter. For instance, the substitution on four letters $\sigma(A) = DAC$, $\sigma(B) = DBABC$, $\sigma(C) = DBAC$, and $\sigma(D) = DABC$ is proper, since all substituted tiles begin with D and end with C.

Proper substitutions force the border (in one step), so we can compute $\check{H}^1$ from the uncollared Anderson-Putnam complex. This complex is typically *not* the wedge of n circles, as an edge need not trace out a closed loop in Γ_0. However, a proper substitution takes each edge to a closed loop, as all vertices in Γ_0 are mapped to a single point. Since $\partial \circ \sigma$ is zero on edges, $\sigma^* \circ \delta$ is zero on 0-cochains. This means that the direct limit under σ^* of $H^1(\Gamma) = C^1/\delta(C^0)$ is the same as the direct limit of $C^1 = \mathbb{Z}^n$ under the dual of the substitution map on edges. However, the dual of the substitution map on edges is just M^T. This proves:

THEOREM 6.1. *If σ is a proper, primitive and aperiodic substitution on n letters, then $\check{H}^1(\Omega_\sigma)$ is the direct limit of $\mathbb{Z}^d$ under the map M^T.*

In the above example, $M^T = \begin{pmatrix} 1 & 0 & 1 & 1 \\ 1 & 2 & 1 & 1 \\ 1 & 1 & 1 & 1 \\ 1 & 1 & 1 & 1 \end{pmatrix}$, which has rank 2 and nonzero eigenvalues 1 and 4, with eigenvectors $(1, -1, 0, 0)^T$ and $(2, 4, 3, 3)^T$. The direct limit is $\mathbb{Z}[1/4] + \mathbb{Z}$.

EXERCISE 6.1. *Show that a substitution forces the border if a power of the substitution is proper.*

EXERCISE 6.2. *Show that a substitution on two letters that forces the border must have a power that is proper.*

Rewriting a substitution. Most of the one-dimensional substitutions we have seen in this book are not proper. However a power of a primitive substitution can always be rewritten in a way that makes it proper [**Dur**].

If σ is a primitive substitution on n letters, then there are at most n^2 possible vertex types – n choices for the tile before the vertex and n choices for the tile after. The substitution maps this finite vertex set to itself, and so for any vertex v_0, the sequence of points $\sigma^m(v_0)$ is eventually periodic. That means there exists a vertex v_1 and a positive integer $k \le n^2$ such that $\sigma^k(v_1) = v_1$. For instance, in the Thue-Morse substitution $a \to ab$, $b \to ba$, the vertex $b.a$ (meaning the point where the end of a b tile meets the beginning of an a tile) is mapped to itself by σ^2. Replacing σ with σ^k does not change the tiling space, so we can assume without loss of generality that that there is a vertex $v = x.y$ (where x might equal y) with $\sigma(v) = v$. Put another way, there are tiles x and y such that $\sigma(x)$ ends in x and $\sigma(y)$ begins with y.

Now consider possible patches of a tiling in Ω_σ. There are only a finite number of possible words between successive occurences of the vertex v. These words all begin with y and end in x, and are preceded by x and followed by y. From the patch

$$abb.ab.aabb.aab.abb.ab.aab.abb.aabb.ab.aab \qquad (6.1)$$

we see that for the Thue-Morse substitution, these words are $A = (b)ab(a)$, $B = (b)abb(a)$, $C = (b)aab(a)$ and $D = (b)abb(a)$. As usual, the letters in parentheses denote what comes immediately before and after the given word, and are not part of the word.

Since (squared) substitution maps the vertex v to itself, it maps a word from v to v into a longer word from v to v, and hence maps A, B, C, and D into (finite) combinations of A, B, C, and D. Since all four words A, B, C, D start with a, they all map to words starting with $abba$, hence starting with D. Likewise, they all map to words ending in C. In particular, A goes to DAC, B to $DBABC$, C to $DBAC$ and D to $DABC$. Written in terms of the naive letters a and b, the Thue-Morse substitution was not proper, and no power of the substitution is proper. Rewritten in terms of A, B, C, and D, the square of the Thue-Morse substitution is proper.

The same construction works for *any* primitive substitution. Find a vertex $v = x.y$ and a power k such that $\sigma^k(v) = v$. Replace σ with $\sigma' = \sigma^k$. Write down all of the words that occur between successive occurences of v. These are our new alphabet, and map to combinations of themselves. The substitution σ' on our new alphabet isn't necessarily proper, but a power of σ' necessarily is.

EXERCISE 6.3. *Prove this last statement. Use the fact that every element of the new alphabet begins with y and ends with x.*

EXERCISE 6.4. *Rewrite the Thue-Morse substitution, or rather its square, using the vertex a.a. How does the rewritten substitution compare to the one obtained from b.a?*

In the Thue-Morse substitution, there were more letters in the rewritten substitution than in the original substitution. This is not always the case. If you rewrite the square of the Fibonacci substitution (i.e., $a \to ab$, $b \to bab$) using the fixed vertex $b.a$, then there are only two words $A = ab$ and $B = abb$, with $\sigma'(A) = BA$ and $\sigma'(B) = BBA$.

The difference between Thue-Morse and Fibonacci can be understood in terms of cohomology. Since H^1 of the Thue-Morse space is not the direct limit of the transpose of the substitution matrix, the rewritten substitution must have a matrix that is qualitatively different from the original. The cohomology of the Fibonacci space *is* the direct limit of the transpose of the substitution matrix, so it is not surprising (although hardly obvious) that the rewritten substitution has the same matrix as the square of the original substitution, and in particular has the same number of letters.

Starting and stopping rules. In dynamical systems we frequently look at a recurrent set, meaning a set that is hit by every orbit infinitely often, and consider the *first return map* from that set to itself. Rewriting a substitution to make it proper can be viewed in this light. Our recurrent set (with respect to the action of translation) is all tilings where a vertex of type v sits at the origin. Under leftward translation, the first return path is essentially the word that occurs between the v vertex at the origin and the next v vertex in the tiling. That is, the alphabet of our rewritten substitution is the set of words associated with first return paths.

The magic of the rewriting is that the substitution sends these words to combinations of themselves. This is turn has to do with the substitution sending v to itself. However, we don't need an invariant vertex to make this idea work. We just need an invariant set of vertices.

Let S be a nonempty set of vertices with the property that σ of every element of S is an element of S. There are many possibilities. S could even be the set of *all* vertices! Less trivially, if $\sigma(a)$ begins with a, then S could be the set of vertices of the form $x.a$.

Let $\mathcal{A}$ be the set of all words that occurs between successive elements of S. Since the substitution maps S to itself, it also converts each element of $\mathcal{A}$ to a string of elements of $\mathcal{A}$. That is, we can rewrite our substitution using the new alphabet $\mathcal{A}$. The elements of S are called *starting and stopping rules*.

For example, suppose that σ is a substitution on three letters: $\sigma(a) = bc$, $\sigma(b) = a$, $\sigma(c) = bc$. The following vertices occur in the tiling: $a.b$, $b.c$, $c.a$, and $c.b$. Let $S = \{a.b, c.a, c.b\}$. The only words that occur between elements of S are $A = a$ and $B = bc$, and our substitution reduces to $\sigma(A) = B$, $\sigma(B) = AB$. Our original substitution was a disguised version of Fibonacci, with the B tile split into two pieces. By rewriting, we rejoined the pieces.

Note that rewriting a substitution with a general set S doesn't always make it proper. As an extreme example, if S is the set of all vertices, then $\mathcal{A}$ is just the original alphabet, and the rewritten substitution is identical to the original substitution.

One can generalize starting and stopping rules further, looking for words that occur between more complicated patterns. However, that is equivalent to first working with collared tiles (or n-fold collared tiles) and picking an invariant set of vertices between collared tiles.

6.2. Partial collaring

In the Anderson-Putnam construction, we use uncollared tiles if the substitution forces the border and collared tiles if it doesn't. However, using collared tiles is often overkill. The Fibonacci substitution $a \to b$, $b \to ab$ doesn't force the border, but all supertiles end in b, so there is nothing gained by collaring tiles on the left.

There is a systematic procedure for collaring just enough. Let σ be a primitive and aperiodic substitution, and let t_1 and t_2 be tiles in tilings $T_1, T_2 \in \Omega_\sigma$. We say that t_1 and t_2 are equivalent if, for some integer n, the patch in $\sigma^n(T_1)$ obtained from $\sigma^n(t_1)$ and its nearest neighbors is a translate of the patch in $\sigma^n(T_2)$ obtained from $\sigma^n(t_2)$ and its nearest neighbors. Equivalence classes are called *partially collared tiles*.

If the original substitution forces the border, then tiles of the same type are equivalent, partially collared tiles are just ordinary tiles. (Strictly speaking, collared tiles may be equivalence classes of ordinary tiles, since two ordinary tiles may have the same image after substitution. This does not happen in the examples considered in this book.) If the original substitution does not force the border, then partially collared tiles provide just enough additional information to force the border.

EXERCISE 6.5. *Prove that a substitution written in terms of partially collared tiles necessarily forces the border.*

In the Fibonacci tiling, there are four kinds of once-collared tiles, namely $(b)a(b)$, $(a)b(a)$, $(a)b(b)$ and $(b)b(a)$. However, $(a)b(a)$ and $(b)b(a)$ both substitute to $(b)ab(b)$, so there are only three partially-collared tiles: $A = a(b)$, $B = b(a)$ and $C = b(b)$, and the substitution is $\sigma(A) = B$, $\sigma(B) = AC$, $\sigma(C) = AB$. The Anderson-Putnam complex is shown below. H^1 of this complex is $\mathbb{Z}^2$, and substitution induces a map $\left(\begin{smallmatrix} 1 & 1 \\ 1 & 0 \end{smallmatrix}\right)$ on this $\mathbb{Z}^2$, so $H^1(\Omega)$ is the direct limit of $\left(\begin{smallmatrix} 1 & 1 \\ 1 & 0 \end{smallmatrix}\right)$, namely $\mathbb{Z} \oplus \tau\mathbb{Z}$.

The same idea works in higher dimensions. In the arrow version of the chair tiling, there are 13 collared tiles, up to rotation. However, there are only six partially collared tiles (up to rotation). If a tile has an arrow pointing northeast, then the neighboring tiles to the northwest and southeast are irrelevant, since after one substitution we always get tiles pointing northwest and southeast, respectively, in those positions. Only the tiles that touch the head of the arrow (i.e., the northern, eastern and northeastern neighbors),

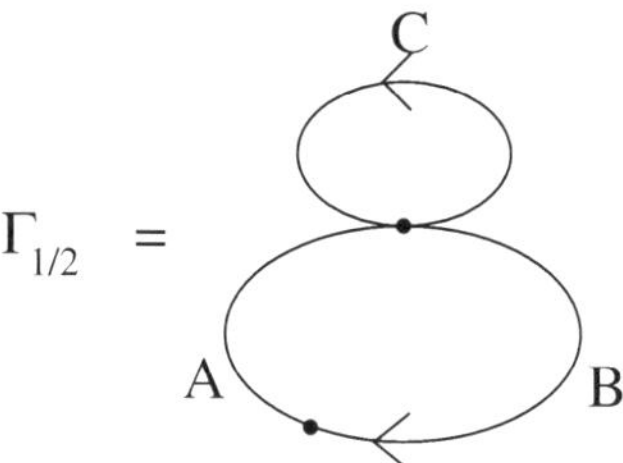

FIGURE 6.1. The Fibonacci tiling's partially collared complex

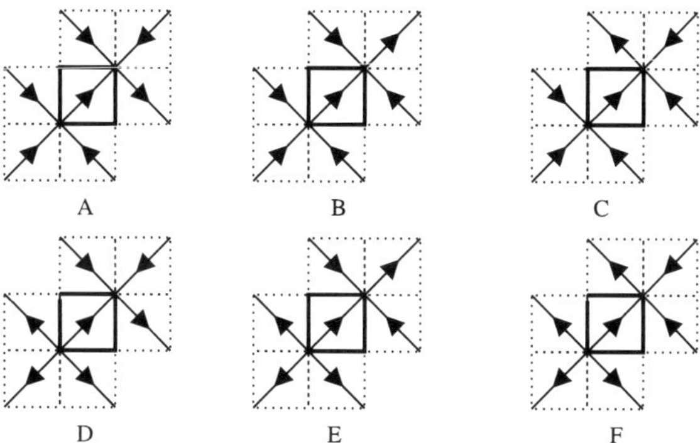

FIGURE 6.2. Six partially collared arrow tiles

or the tail of the arrow (southern, western and southwestern neighbors), matter. There are two possible pattern at the tail (three arrows pointing in and one out, or all four pointing out), and three possible patterns at the head (exactly one of the three other arrows is pointing out).

In Chapter 4 we did a "flat earth" calculation of the cohomology of the arrow tiling using uncollared tiles. The following exercises walk you through the "round earth" calculation using partially collared tiles.

We denote the six partially collared tile types, in standard orientation, as A, B, C, D, E, and F, as in Figure 6.2, and let the prefix r denote a counterclockwise rotation by $\pi/2$.

EXERCISE 6.6. *Write down the substitution rule in terms of partially collared tiles. Note that each tile in standard orientation gives rise to two tiles in standard orientation, one tile rotated by $\pi/2$ and one tile rotated by $-\pi/2$.*

EXERCISE 6.7. *From the adjacencies within the order-2 supertiles, deduce edge and vertex identifications. (There are no others.) Make a complete list of edges and vertices. Note that, if v is a vertex, then v, rv, r^2v and r^3v may not all be different. Decompose your list into representations of the rotation group.*

EXERCISE 6.8. *Write down the transformations ∂_2 and ∂_1, as matrices whose entries are polynomials in r. The matrices for δ_1 and δ_0 are the transposes of the matrices for ∂_2 and ∂_1, respectively, only with r replaced by r^{-1}.*

EXERCISE 6.9. *One representation at a time, compute the cokernel of δ_1, the kernel of δ_1 and the image of δ_0. From these, deduce the cohomology of the partially-collared Anderson-Putnam complex.*

EXERCISE 6.10. *See how each term you computed in the previous exercise transforms under substitution. Take the direct limit to get the cohomology of the chair tiling space.*

EXERCISE 6.11. *Compare to the results of the "flat earth" calculation. Show that the generators of the $\mathbb{Z}[1/2]$ terms (i.e., the contributions of the $r^2 = -1$ representation) can be written as cochains that are pullbacks on cochains on the uncollared complex, as can 3 times the generator of the rotationally invariant part of $\check{H}^2$, but that the generator itself requires collared tiles.*

6.3. Relative cohomology and eventual ranges

So far we have worked exclusively with absolute cohomology. There are many instances where it is helpful to break our tiling space (or its approximants) into pieces and use relative homology.

Relative homology and cohomology for CW complexes. Let X be a nice topological space, such as a CW complex, and let Y be a subspace of X. We can consider the space $C_k(Y)$ of k-chains on Y, which is a subgroup of $C_k(X)$. (It doesn't matter whether we're talking about simplicial chains, singular chains, or cellular chains, as the construction is the same in each category.) The *relative chain group* $C_k(X, Y)$ is just the quotient $C_k(X)/C_k(Y)$. Put another way, we have an exact sequence

$$0 \to C_k(Y) \to C_k(X) \to C_k(X, Y) \to 0, \tag{6.2}$$

where the first nontrivial map is inclusion and the last is a quotient. If two chains on X differ by a chain on Y, then their boundaries also differ by a chain on Y. This means that the boundary map can be viewed as a map $\partial_k : C_k(X, Y) \to C_{k-1}(X, Y)$ on relative chain groups, and $\partial^2 = 0$ as usual.

The *relative homology group* $H_k(X, Y)$ is the kernel of ∂_k modulo the image of ∂_{k+1}. If (X, Y) is a "good pair", (e.g., X is a CW complex and Y is a subcomplex,) then $H_k(X, Y)$ is isomorphic to the k-th reduced homology of the quotient space X/Y. For a proof of this fact, see [**Hat**].

By the snake lemma (again, see [**Hat**]), the exact sequence of chain complexes (6.2) induces a exact sequence of homology groups

$$\cdots \to H_k(Y) \to H_k(X) \to H_k(X, Y) \to H_{k-1}(Y) \to \cdots, \tag{6.3}$$

called the *homology long exact sequence of the pair* (X, Y). Using this long exact sequence, we can frequently deduce the homology of a big space X from the homologies of Y and X/Y.

Cohomology is similar. The relative cochain group $C^k(X, Y)$ is defined to be the dual of $C_k(X, Y)$. That is, it is the set of cochains on X that evaluate to zero on chains on Y.

EXERCISE 6.12. *Show that the coboundary operator* $\delta_k = \partial^*_{k+1}$ *maps* $C^k(X, Y)$ *to* $C^{k+1}(X, Y)$ *and that* $\delta^2 = 0$.

We then define the *relative cohomology groups* $H^k(X, Y)$ to be the quotients $\ker \delta_k / \operatorname{Im} \delta_{k-1}$, where each δ is viewed as a map on relative cochains.

Each relative cochain is also a cochain on X, and every cochain on X can be restricted to Y, giving us a short exact sequence of cochain complexes

$$0 \to C^k(X, Y) \to C^k(X) \to C^k(Y) \to 0, \tag{6.4}$$

which induces the *cohomology long exact sequence of the pair* (X, Y)

$$\cdots \to H^k(X, Y) \to H^k(X) \to H^k(Y) \to H^{k+1}(X, Y) \to \cdots . \tag{6.5}$$

There is also a long exact sequence in reduced cohomology,

$$\cdots \to H^k(X, Y) \to \tilde{H}^k(X) \to \tilde{H}^k(Y) \to H^{k+1}(X, Y) \to \cdots , \tag{6.6}$$

where the *reduced cohomology groups* are $\tilde{H}^k(X) = H^k(X)$ for $k > 0$ and $\tilde{H}^0(X) = H^0(X)/\mathbb{Z}$.

Inverse limit spaces and Čech cohomology. Unfortunately, tiling spaces are not CW complexes, so the above ideas cannot be applied directly to them. Simplicial, singular and cellular cohomology do not work (in any reasonable way) on tiling spaces! Fortunately, the Čech cohomology of a tiling space *is* well defined, as we saw in Chapter 3.

If X contains Y, then every open cover of X can be restricted to Y, and every open cover of Y can be extended to an open cover of X. If $\mathcal{U}$ is an open cover of X and $\mathcal{U}|_Y$ is its restriction to Y, then the nerve of $\mathcal{U}_Y$ is a sub-complex of the nerve of $\mathcal{U}$. Since a Čech cochain on the open cover $\mathcal{U}$ is just a cellular cochain on the nerve of $\mathcal{U}$, we can repeat the construction of relative cohomology for CW complexes to get the relative Čech cohomology of an open cover. Taking the direct limit over all open covers, we get the relative cohomology of an arbitrary pair (X, Y) with $Y \subset X$.

We don't really want to work with arbitrary open covers if we can help it. If we view a tiling space as an inverse limit of CW complexes, then it's easier to view the Čech cohomology of the tiling space as the direct limit of the (Čech , singular or cellular) cohomology of the approximants. We can do the same for relative Čech cohomology.

THEOREM 6.2. *Let* $X = \varprojlim(X_n, f_n)$, *where each* X_n *is a CW complex. Let* Y_n *be a subcomplex of* X_n, *and suppose that* $f_n(Y_n) \subset Y_{n-1}$. *Let* $Y =$

$\varprojlim(Y_n, f_n)$. *There is a long exact sequence*

$$\cdots \to \check{H}^k(X, Y) \to \check{H}^k(X) \to \check{H}^k(Y) \to \check{H}^{k+1}(X, Y) \to \cdots \qquad (6.7)$$

among the Čech cohomologies of the inverse limits spaces. Furthermore each term in the long exact sequence, and the maps between them, are direct limits of the corresponding terms in the cohomology long exact sequences of the approximants. This theorem also applies to reduced cohomology.

PROOF. We previously proved that $\check{H}^k(X)$ is the direct limit of $H^k(X_n)$. That proof involved considering a cofinal set of open covers of X obtained from progressively finer good open covers of approximants. The same proof works for relative cohomology, as long as we pick our good open covers of X_n such that their restrictions to Y_n are also good. If $\mathcal{U}$ is such a good cover, then the relative Čech cohomology of $\mathcal{U}$ is isomorphic to the (ordinary) relative cohomology of the nerves of $\mathcal{U}$ and $\mathcal{U}|_{Y_n}$, which is isomorphic to the (ordinary) relative cohomology $H^k(X_n, Y_n)$. Taking the limit over open covers, we get that $\check{H}^k(X, Y)$ is the direct limit of $H^k(X_n, Y_n)$.

The exactness of the sequence (6.7) comes from the naturality of the long exact sequences of (X_n, Y_n) under the maps f_n^*. That is, the fact that the diagram

$$
\begin{array}{ccccccc}
H^k(X_{n-1}, Y_{n-1}) & \to & H^k(X_{n-1}) & \to & H^k(Y_{n-1}) & \to & H^{k+1}(X_{n-1}, Y_{n-1}) \\
\downarrow{\scriptstyle f_n^*} & & \downarrow{\scriptstyle f_n^*} & & \downarrow{\scriptstyle f_n^*} & & \downarrow{\scriptstyle f_n^*} \\
H^k(X_n, Y_n) & \to & H^k(X_n) & \to & H^k(Y_n) & \to & H^{k+1}(X_n, Y_n)
\end{array}
\qquad (6.8)
$$

commutes. Let β be an element of a group in (6.7), let i be the map to that group, and let j be the map from that group. If $\beta = i(\alpha)$ for some α in the previous group, then α can be represented by an element α_n in $H^k(X_n)$ or $H^k(Y_n)$ or $H^k(X_n, Y_n)$, so β is represented by $i_n(\alpha_n)$, so $j(\beta)$ is represented by $j_n(i_n(\alpha_n)) = 0$. Conversely, if $j(\beta) = 0$, then β is represented by some β_n with $j_n(\beta_n) = 0$, so $\beta_n = i_n(\alpha_n)$, so $\beta = i(\alpha)$, where α is represented by α_n. $\qquad\square$

Eventual ranges. Let W be a finite set, and let $f : W \to W$ be a map. By assumption, $f(W) \subset W$, and so $f(f(W)) \subset f(W)$, and generally $f^{k+1}(W) \subset f^k(W)$. Since W is finite, the inclusions can't all be proper, and there must be an N for which $f^{N+1}(W) = f^N(W)$. But then $f^{N+2}(W) = f^N(W)$, etc., and hence for all $n \geq N$, $f^n(W) = f^N(W)$. The subset $f^N(W)$ is called the *eventual range* of the map f.

The same idea works for linear maps on finite-dimensional vector spaces. If V is a finite-dimensional vector space and $L : V \to V$ is a linear map, then $L(V)$ is a subspace of V, $L^2(V)$ is a subspace of $L(V)$, and so on. After a finite number of steps we have $L^{N+1}(V) = L^N(V)$, and we call $L^N(V)$ the *eventual range* of L.

Finally, and most importantly for our purposes, the same idea applies to substitution maps on Anderson-Putnam complexes, and on their lower-dimensional skeleta. If Γ is a finite CW complex and σ is a cellular map, then $\sigma(\Gamma)$ is a subcomplex of Γ, and the sub-complexes $\sigma^k(\Gamma)$ stabilize to an eventual range, which we denote Γ_{ER}.

The key algebraic facts about eventual ranges are:

THEOREM 6.3. *Let Γ be a CW complex and σ a cellular map. Let Γ_0 be a subcomplex of Γ such that $\sigma(\Gamma_0) \subset \Gamma_0$. Then*

(1) $\varprojlim(\Gamma, \sigma) = \varprojlim(\Gamma_{ER}, \sigma)$
(2) $\check{H}^k(\varprojlim(\Gamma, \sigma)) = \varinjlim H^k(\Gamma_{ER})$
(3) $\check{H}^k(\varprojlim(\Gamma, \sigma), \varprojlim(\Gamma_0, \sigma)) = \varinjlim H^k(\Gamma_{ER}, (\Gamma_0)_{ER}).$
(4) $\check{H}^k(\varprojlim(\Gamma, \sigma), \varprojlim(\Gamma_0, \sigma)) = \varinjlim H^k(\Gamma_{ER}, \Gamma_0 \cap \Gamma_{ER}).$

PROOF. Let N and N_0 be such that $\sigma^N(\Gamma) = \Gamma_{ER}$ and $\sigma^{N_0}(\Gamma_0) = (\Gamma_0)_{ER}$. If $(x_0, x_1, \ldots)$ is a sequence of points in Γ, with each $x_i = \sigma(x_{i+1})$, then each $x_i = \sigma^N(x_{i+N}) \in \Gamma_{ER}$, so every point in the inverse limit of Γ is actually in the inverse limit of Γ_{ER}. This proves the first claim, and the second and third claims follow immediately from the first. The subtlety is in the fourth, where we use the eventual range of Γ but do not use the eventual range of Γ_0. Suppose that $\alpha \in C^k(\Gamma_{ER}, (\Gamma_0)_{ER})$. That is, α is a k-cochain that vanishes on chains supported in $(\Gamma_0)_{ER}$. However, if c is a chain on Γ_0, then $\sigma^{N_0}(c)$ is a chain on $(\Gamma_0)_{ER}$, so $((\sigma^*)^{N_0}\alpha)(c) = \alpha(\sigma^{N_0}(c)) = 0$. Thus the direct limit of $C^k(\Gamma_{ER}, (\Gamma_0)_{ER})$ is the direct limit of $C^k(\Gamma_{ER}, \Gamma_{ER} \cap \Gamma_0)$. The relative cohomology of the inverse limit is computed from the action of δ on this direct limit, and so can be computed either from the direct limit of $H^k(\Gamma_{ER}, (\Gamma_0)_{ER})$ or from the direct limit of $H^k(\Gamma_{ER}, \Gamma_{ER} \cap \Gamma_0)$. $\square$

6.4. Barge-Diamond collaring

All the collaring techniques so far have involved entire tiles. Barge and Diamond [**BD**] take a different approach. Their essential idea is to attach to each point information about the tiling in a small neighborhood of that *point*. If the point is in the interior of a tile, not too close to the boundary, then there is no additional information to be added. If it is close to an edge (but not to a vertex), then the added information is the neighboring tile across the edge. If our point is close to a vertex, our added information describes the tiles meeting at that vertex. The resulting complex has different pieces representing possible (uncollared) tiles, possible edges, and possible vertices. The Barge-Diamond (BD) complex is typically much simpler than the once-collared Anderson-Putnam complex, or even the partially collared Anderson-Putnam complex. It also allows us to understand the different components of $\check{H}^*(\Omega)$. Some come from population statistics (as encoded in the matrix of the substitution), some come from the combinatorics of how tiles meet, and some involve both.

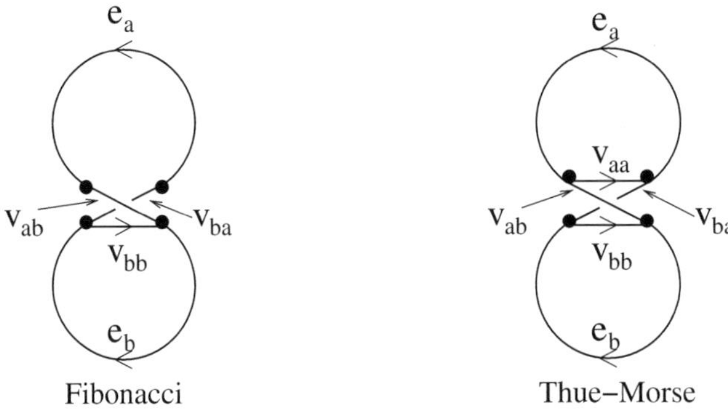

FIGURE 6.3. Two Barge-Diamond complexes

One dimensional substitutions. We begin, as usual, in one dimension. Suppose we have a substitution on a set of tiles $\{t_1, t_2, \ldots, t_n\}$, with t_i having length L_i. Pick an ϵ smaller than half of the smallest L_i. Identify points in a tiling if their local patterns agree out to distance ϵ. If p_1 and p_2 are corresponding points in two tiles of the same type, and if they are not within ϵ of the boundary of their tiles, then they are identified. If they are within ϵ of the left edge, then they are identified only if they have the same neighboring tile to the left; if they are within ϵ of the right edge, they are identified only if they have the same neighboring tile to the right.

The quotient of the tiling by this identification is a complex with edges e_i of length $L_i - 2\epsilon$ representing the interior of tiles a_i, and edges v_{ij} of length 2ϵ representing the vertices $a_i.a_j$. We call the e_i's *tile cells* and the v_{ij}'s *vertex flaps*. The left endpoint of v_{ij} is the right endpoint of e_i, and the right endpoint of v_{ij} is the left endpoint of e_j, so in particular the left endpoints of v_{ij} and v_{ik} are identified. The resulting complex is denoted S, and the sub-complex of just the edge flaps is denoted S_0.

For example, for the Fibonacci tiling, we have tile cells e_a and e_b and vertex flaps v_{ab}, v_{ba} and v_{bb}. In the Thue-Morse tiling, we have tile cells e_a and e_b and vertex flaps v_{ab}, v_{ba}, v_{aa} and v_{bb}. Although Fibonacci and Morse have the same uncollared Anderson-Putnam complexes (a figure 8), they have different Barge-Diamond complexes. The BD complex for Fibonacci has the homotopy type of a figure 8, but the BD complex for Thue-Morse has an extra loop, comprised entirely of edge flaps. As we shall see, this extra loop accounts for a factor of $\mathbb{Z}$ in H^1 of the Thue-Morse tiling space.

If the points p and q are identified in a tiling T, then their neighborhoods agree to distance ϵ, so the neighborhoods of λp and λq in $\sigma(T)$ agree to distance $\lambda\epsilon$, where λ is the stretching factor of the substitution σ. $\lambda\epsilon > \epsilon$, so λp and λq are identified in $\sigma(T)$. Thus the substitution induces a map (also denoted σ) from S to itself.

THEOREM 6.4. $\varprojlim(S, \sigma) = \Omega_\sigma$.

PROOF. A point in the 0th approximant describes the part of a tiling the lies within ϵ of the origin. A point in the nth approximant describes the tiling out to a distance $\lambda^n \epsilon$. A point in the inverse limit is a consistent description of the tiling out to infinite distance, and so is identified with the tiling itself. $\square$

This theorem tells us how to compute the cohomology of a tiling space. First construct the BD complex and compute its cohomology. For instance, Fibonacci has $H^1(S) = \mathbb{Z}^2$ and Thue-Morse has $H^1(S) = \mathbb{Z}^3$. Then take the direct limit of that cohomology under the map σ^*. The trouble is that σ is not a cellular map. It takes a vertex flap to an interval of length $2\lambda\epsilon$, which is a vertex flap plus parts of two tile cells. This makes it hard to see what σ^* does. The solution is to find a cellular map $\bar{\sigma}$ that is homotopic to σ, so that $\bar{\sigma}^* = \sigma^*$ as a map in cohomology. For any point $p \in S$, $\bar{\sigma}(p)$ is obtained from $\sigma(p)$ by flowing to the center of the nearest vertex flap in such a way that intervals of size $2\lambda\epsilon$ around the center of an vertex flap become intervals of size 2ϵ. (The exact formulas are not important, but you can use a linear contraction on these intervals, coupled with a linear expansion of each tile cell about its center.) In this way, each vertex flap maps to a single vertex flap, and each tile cell maps to a union of tile cells and vertex flaps.

Since the complex S_0 maps to itself, we can consider the inverse limit $\Xi_0 = \varprojlim(S_0, \bar{\sigma})$ as a subset of $\Xi = \varprojlim(S, \bar{\sigma})$. We then use the (reduced) cohomology exact sequence of the pair (Ξ, Ξ_0)

$$0 \to \check{H}^0(\Xi, \Xi_0) \quad \to \quad \check{H}^0(\Xi) \quad \to \quad \check{H}^0(\Xi_0)$$
$$\hookrightarrow \quad \check{H}^1(\Xi, \Xi_0) \quad \to \quad \check{H}^1(\Xi) \quad \to \quad \check{H}^1(\Xi_0) \to 0. \tag{6.9}$$

to get $\check{H}^*(\Omega) = \check{H}^*(\Xi)$ from $\check{H}^*(\Xi_0)$ and the relative cohomology groups $\check{H}^*(\Xi, \Xi_0)$. Since Ξ is connected, the first two terms vanish and we have

$$0 \to \tilde{H}^0(\Xi_0) \to \check{H}^1(\Xi, \Xi_0) \to \check{H}^1(\Xi) \to \check{H}^1(\Xi_0) \to 0. \tag{6.10}$$

The quotient space S/S_0 is a wedge of circles, one for each tile type, and $H^1(S, S_0)$ is free, with rank equal to the number of letters in our alphabet. The action of $\bar{\sigma}'$ on $H^1(S, S_0)$ is just the transpose of the substitution matrix M, and $\check{H}^1(\Xi, \Xi_0)$ is the direct limit of M^T.

What remains is to understand $\tilde{H}^0(\Xi_0)$ and $\check{H}^1(\Xi_0)$ and to apply equation (6.10). For the Fibonacci tiling this is easy, since S_0 is contractible. In this case equation (6.9) says that $\check{H}^1(\Xi) = \check{H}^1(\Xi, \Xi_0) = \lim M^T = \mathbb{Z}^2$.

For the Thue-Morse tiling, S_0 is a closed loop, so $\tilde{H}^0(S_0) = 0$ and $H^1(S_0) = \mathbb{Z}$. Substitution merely reverses the direction of the loop, so $H^1(\Xi_0) = \mathbb{Z}$. Since $\check{H}^1(\Xi, \Xi_0) = \lim M^T = \mathbb{Z}[1/2]$, equation (6.10) reads:

$$0 \to 0 \to \mathbb{Z}[1/2] \to \check{H}^1(\Xi) \to \mathbb{Z} \to 0, \tag{6.11}$$

from which we get $\check{H}^1(\Xi) = \mathbb{Z} \oplus \mathbb{Z}[1/2]$.

These results are not new – we computed these examples before – but they give a new insight into the two terms in $\check{H}^1$ of the Thue-Morse tiling space. The $\mathbb{Z}[1/2]$ term counts tiles, or supertiles of various orders, and that is described by the substitution matrix. The $\mathbb{Z}$ term comes from the combinatorics of how adjacent tiles can meet, and that comes from S_0.

To understand one-dimensional substitutions in general, we look at the eventual range of S_0.

THEOREM 6.5 (Barge-Diamond). *Suppose that the eventual range $(S_0)_{ER}$ has k connected components and ℓ independent loops. Then*

$$0 \to \mathbb{Z}^{k-1} \to \lim M^T \to \check{H}^1(\Omega) \to \mathbb{Z}^\ell \to 0 \qquad (6.12)$$

is exact, where M is the substitution matrix. If $k = 1$, then the sequence splits, and $\check{H}^1(\Omega) = \mathbb{Z}^\ell \oplus \lim M^T$. If $\ell = 0$, then $\check{H}^1(\Omega)$ is the quotient of $\lim M^T$ by $\mathbb{Z}^{k-1}$.

PROOF. The tiling space Ω has the same cohomology as the inverse limit Ξ, whose cohomology we compute from (6.10). Substitution merely permutes vertex flaps in the eventual range, so σ^* is an isomorphism on $H^*((S_0)_{ER})$, so $\check{H}^*(\Xi_0) = H^*((S_0)_{ER})$. In particular, $\tilde{H}^0(\Xi_0) = \mathbb{Z}^{k-1}$ and $\check{H}^1(\Xi_0) = \mathbb{Z}^\ell$, which gives (6.12). If $k = 1$, then we have a short exact sequence whose final term is free, so the sequence splits. If $\ell = 0$, then we have a short exact sequence that determines $\check{H}^1(\Omega)$. In general, however, it is not possible to deduce $\check{H}^1(\Omega)$ entirely from k, ℓ and M^T. You need to know the maps in the exact sequence (6.12). $\square$

COROLLARY 6.6. *If $(S_0)_{ER}$ is contractible, then $\check{H}^1(\Omega) = \lim M^T$.*

We have already seen examples of this. If a substitution is proper, then $(S_0)_{ER}$ consists of a single vertex flap. More generally, if all substituted letters end with the same letter, as with the Fibonacci substitution, then all the vertex flaps in the eventual range are of the form $x.y$, with x fixed, and $(S_0)_{ER}$ retracts to the shared left endpoint of all these vertex flaps. Of course, the situation where all substituted letters *begin* with the same letter is similar.

Tilings in higher dimensions. For simplicity, we will talk about 2-dimensional tilings, but similar ideas apply in more than two dimensions. Our working example will be a tiling by three kinds of unit square tiles, with the substitution rule

$$\sigma(A) = \begin{matrix} A & B \\ A & C \end{matrix} \qquad \sigma(B) = \begin{matrix} A & B \\ B & C \end{matrix} \qquad \sigma(C) = \begin{matrix} A & B \\ C & C \end{matrix}. \qquad (6.13)$$

This substitution does not force the border, and the fully collared Anderson-Putnam complex is horrific. Even the partially collared complex is cumbersome. The Barge-Diamond complex, however, is manageable.

Pick an $\epsilon < 1/2$, and identify points in a tiling if their square neighborhoods, of side 2ϵ, agree up to translation. If p_1 and p_2 are corresponding points in two tiles of the same type, and are not within ϵ of an edge, then they are identified. If they are within ϵ of an edge, but not within ϵ of two edges, then they are identified if their neighboring tiles (across the edge) agree. If they are within (ϵ, ϵ) of a vertex, then the pattern of tiles around that vertex must agree.

Instead of having tile cells and vertex flaps, we now have *three* kinds of cells. We have $(1 - 2\epsilon) \times (1 - 2\epsilon)$ *tile cells* corresponding to the interiors of each tile, $2\epsilon \times (1 - 2\epsilon)$ *edge flaps* that describe all the possible edges (away from the vertices), and $2\epsilon \times 2\epsilon$ *vertex squares*. Let S_0 be the union of the vertex squares, let S_1 be the union of the vertex squares and edge flaps, and let $S = S_2$ be the union of all of the cells.

As before, $\Omega = \varprojlim(S, \sigma)$, so $H^*(\Omega)$ is the direct limit under substitution of $H^*(S)$. Also as before, we take a homotopy $\bar{\sigma}$ of our substitution that sends S_0 to S_0 and that sends S_1 to S_1, we let $\Xi_0 = \varprojlim(S_0, \bar{\sigma})$, $\Xi_1 = \varprojlim(S_1, \bar{\sigma})$, and $\Xi = \Xi_2 = \varprojlim(S_2, \bar{\sigma})$, and have

$$\check{H}^*(\Omega) = H^*(\varprojlim S, \sigma) = \varinjlim H^*(S, \sigma^*) = \varinjlim H^*(S, \bar{\sigma}^*) = \check{H}^*(\Xi). \quad (6.14)$$

We compute $\check{H}^*(\Xi)$ in stages. First we study S_0 and its eventual range, and compute $\check{H}^*(\Xi_0)$. We then compute $\check{H}^*(\Xi_1, \Xi_0)$ from the direct limit of $H^*((S_1)_{\mathrm{ER}}, S_0 \cap (S_1)_{\mathrm{ER}})$. Unlike with 1-dimensional tilings, where we always got the direct limit of M^T, this step requires real work. We combine these results using the cohomology long exact sequence of the pair (Ξ_1, Ξ_0) to get $\check{H}^*(\Xi_1)$. Then we compute $\check{H}^*(\Xi_2, \Xi_1)$, which is obtained from the substitution matrix. Finally, we use the cohomology long exact sequence of the pair (Ξ_2, Ξ_1) to get $\check{H}^*(\Xi_2)$.

In our example, there are many possible vertices, but only three in the eventual range, namely

$$\frac{C\,|\,A}{B\,|\,A}, \qquad \frac{C\,|\,B}{B\,|\,A}, \qquad \text{and} \qquad \frac{C\,|\,C}{B\,|\,A}. \quad (6.15)$$

The union of these three vertex squares is contractible (just retract to their common southwest corner), so $\tilde{H}^0(\Xi_0) = \check{H}^1(\Xi_0) = \check{H}^2(\Xi_0) = 0$.

By Theorem 6.3, we can compute $\check{H}^*(\Xi_1, \Xi_0)$ from the eventual range of S_1 modulo all of S_0. There are seven kinds of vertical edges in S_1, corresponding to the transitions $B|A$, $C|A$, $C|B$, $C|C$, $A|B$, $A|C$, and $B|C$, but only the first four are in the eventual range — the last three only occur in the interior of substituted tiles. Since the $B|A$ and $C|A$ edge flaps are joined at the right, and since that $C|A$, $C|B$ and $C|C$ edge flaps are joined at the left, the union of the four edge flaps is homotopy equivalent to a vertical line segment whose top and bottom points lie in S_0. Likewise, there are seven horizontal edge flaps, but only four in the eventual range ($\frac{A}{A}$, $\frac{B}{A}$, $\frac{C}{A}$, and $\frac{C}{B}$), whose union is homotopy equivalent to a horizontal line segment with endpoints in S_0. The quotient space $(S_1)_{\mathrm{ER}}/S_0$ is homotopy equivalent to the wedge of two circles, one corresponding to the horizontal edges and one corresponding to the vertical edges. The only nontrivial relative cohomology group is $H^1((S_1)_{\mathrm{ER}}, S_0) = \mathbb{Z}^2$. Substitution doubles each element of this group, since each horizontal edge turns into two horizontal edges, and each vertical edge turns into two vertical edges. The direct limit is $H^1(\Xi_1, \Xi_0) = \mathbb{Z}[1/2]^2$.

Since Ξ_0 is contractible, the cohomology long exact sequence of the pair (Ξ_1, Ξ_0) says that the reduced cohomology of Ξ_1 equals the relative cohomology, i.e.,

$$\tilde{H}^0(\Xi_1) = 0, \qquad \check{H}^1(\Xi_1) = \mathbb{Z}[1/2]^2, \qquad \check{H}^2(\Xi_1) = 0. \tag{6.16}$$

The last ingredient of our calculation is $\check{H}^*(\Xi_2, \Xi_1)$. The quotient space S_2/S_1 is a wedge of three spheres, one for each kind of tile. Its only nontrivial (reduced) homology is $H^2 = \mathbb{Z}^3$. Under substitution, this transforms by $M^T = \left(\begin{smallmatrix} 2 & 1 & 1 \\ 1 & 2 & 1 \\ 1 & 1 & 2 \end{smallmatrix}\right)$, so

$$\begin{aligned} \check{H}^1(\Xi_2, \Xi_1) &= \check{H}^0(\Xi_2, \Xi_1) = 0, \\ \check{H}^2(\Xi_2, \Xi_1) &= \lim M^T = \mathbb{Z}[1/4] + \mathbb{Z}^2. \end{aligned} \tag{6.17}$$

Finally, we use the reduced cohomology long exact sequence of the pair (Ξ_2, Ξ_1).

$$\begin{aligned}
0 &\to \tilde{H}^0(\Xi_2) \to 0 \\
\hookrightarrow \quad 0 &\to \check{H}^1(\Xi_2) \to \mathbb{Z}[1/2]^2 \\
\hookrightarrow \quad \mathbb{Z}[1/4] + \mathbb{Z}^2 &\to \check{H}^2(\Xi_2) \to 0.
\end{aligned} \tag{6.18}$$

The map $\mathbb{Z}[1/2]^2 \to \mathbb{Z}[1/4] + \mathbb{Z}^2$ must be zero, since the source and target scale differently under substitution. As a result,

$$\begin{aligned} \check{H}^1(\Omega) &= \check{H}^1(\Xi_2) = \mathbb{Z}[1/2]^2 \\ \check{H}^2(\Omega) &= \check{H}^2(\Xi_2) = \mathbb{Z}[1/4] + \mathbb{Z}^2. \end{aligned} \tag{6.19}$$

Note how the different terms in the answer come from different ingredients. $\check{H}^2$ comes entirely from the substitution matrix, and its generators are the cochains that count the three kinds of tiles. $\check{H}^1$ comes entirely from S_1. All horizontal edges are equivalent, and all vertical edges are equivalent, so the only interesting cochains are the ones that count vertical or horizontal edges, regardless of type. (This is another way to see that the coboundary map $\mathbb{Z}[1/2]^2 \to \mathbb{Z}[1/4] + \mathbb{Z}^2$ is zero.) Taking the direct limit under substitution, we can count the edges of supertiles of arbitrary order. There are no interesting contributions from S_0, insofar as $(S_0)_{\mathrm{ER}}$ is contractible.

The alert reader may have noticed that (6.19) is the same as the cohomology of the half-hex tiling. In fact, this tiling space is homeomorphic to the half-hex tiling space. Each letter refers to a way to get a whole hexagon from two half-hexagons. Rewriting the half-hex substitution to take whole hexagons to whole hexagons and then shearing the triangular lattice to a square lattice completes the equivalence.

The chair tiling, yet again. The previous example was chosen to be easy, especially in the application of the cohomology long exact sequences. We complete this section by (re)computing the cohomology of the chair

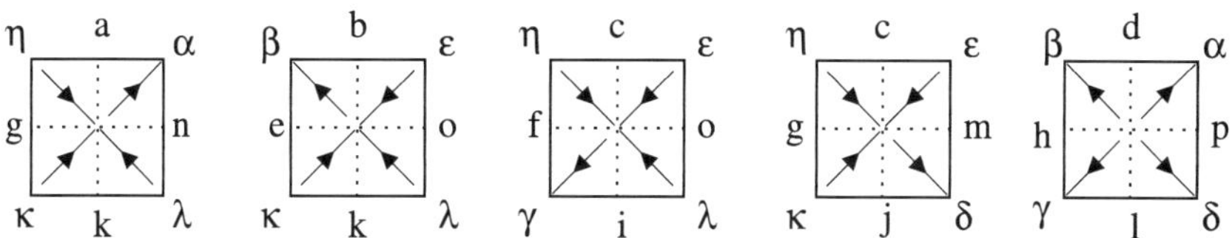

FIGURE 6.4. The 5 vertex squares in $(S_0)_{\mathrm{ER}}$

tiling (as represented by arrows), in full rigor. In this example, the different components interact in a nontrivial way.

For typographical simplicity, we will call a northeast arrow A, a northwest arrow B, a southwest arrow C, and a southeast arrow D. Let f_A, f_B, f_C, and f_D denote the four tile cells. To capture transitions at vertical edges, note that there are eight allowed configurations at vertical edges for which we need flaps, 4 of the form ▱ and 4 of the form ▱, where the double-headed arrow ▱ indicates that either ▱ or ▱ can appear in this position. The edge flap ▱ (denoted e_{AB}) is glued to f_A on the left and f_B on the right, and hence is also glued to e_{AD} on the left and e_{CB} on the right. The four flaps from the configurations ▱ glue together to form a vertical tube, as do the four flaps from ▱. Similarly, there are two distinct horizontal tubes capturing allowed configurations along horizontal edges.

There are two general patterns of allowed configurations at vertices, those of the form ▱ and those of the form ▱. Only the latter appear in the eventual range.

$$(S_0)_{\mathrm{ER}} = \left\{ \;\boxplus\,,\;\boxplus\,,\;\boxplus\,,\;\boxplus\,,\;\boxplus\; \right\}, \qquad (6.20)$$

with substitution taking each of these five vertex squares to itself.

As always, we begin by computing $\check{H}^*(\Xi_0) = H^*((S_0)_{\mathrm{ER}})$. The identifications of the edges and corners are shown in Figure 6.4. After identifications, $(S_0)_{\mathrm{ER}}$ consists of 5 squares, 16 edges and 8 vertices, for an Euler characteristic of -3. It is connected ($H^0 = \mathbb{Z}$) and has trivial H^2, so must have $H^1 = \mathbb{Z}^4$. In fact, it is homotopy equivalent to the wedge of four circles.

As noted earlier, the complex S_1 consists of S_0 together with 16 edge flaps. These edge flaps form four tubes, and $H^1(S_1, S_0) = H^2(S_1, S_0) = \mathbb{Z}^4$. Substitution takes each edge flap of the form ▱ to itself plus ▱, takes ▱ to itself plus ▱, takes ▱ to itself plus ▱, and takes ▱ to itself plus ▱. $\bar{\sigma}^*$ acts by $\begin{pmatrix} 1 & 1 & 0 & 0 \\ 1 & 1 & 0 & 0 \\ 0 & 0 & 1 & 1 \\ 0 & 0 & 1 & 1 \end{pmatrix}$ on $H^1(S_1, S_0)$, but acts trivially on

$H^2(S_1, S_0)$. The upshot is that

$$\check{H}^1(\Xi_1, \Xi_0) = \mathbb{Z}[1/2]^2, \qquad \check{H}^2(\Xi_1, \Xi_0) = \mathbb{Z}^4. \tag{6.21}$$

One generator of $\check{H}^1$ counts horizontal edges of supertiles of arbitrary order, while the other counts vertical edges. The generators of $\check{H}^2$ count the four tubes.

The cohomology long exact sequence of the pair (Ξ_1, Ξ_0) reads:

$$0 \to \mathbb{Z}^4 \to \check{H}^1(\Xi_1) \to \mathbb{Z}^4 \xrightarrow{\delta} \mathbb{Z}^4 \to \check{H}^2(\Xi_1) \to 0. \tag{6.22}$$

To get δ, we compute the boundary of each tube and view it as a chain in S_0, hence a class in $H_1(S_0)$. δ is the dual of the resulting map from $H_2(S_1, S_0)$ to $H_1(S_0)$. The matrix of δ is nonsingular with determinant 3, so $\check{H}^1(\Xi_1) = \check{H}^1(\Xi_1, \Xi_0) = \mathbb{Z}^4$ and $\check{H}^2(\Xi_1) = \mathbb{Z}_3$.

EXERCISE 6.13. *Compute the matrix of δ.*

The quotient space S_2/S_1 is just the wedge of four spheres. Thus $H^1(S_2, S_1) = 0$ and $H^2(S_2, S_1) = \mathbb{Z}^4$. Under substitution, this gets multiplied by the transpose of the substitution matrix, and the limit is $\mathbb{Z}[1/4] \oplus \mathbb{Z}[1/2]^2$. The generator of $\mathbb{Z}[1/4]$ simply counts tiles, regardless of type, while the generators of $\mathbb{Z}[1/2]^2$ count the vector sum of all the arrows.

Finally, we use the (reduced) cohomology exact sequence of the pair (Ξ_2, Ξ_1), which reads

$$0 \to \check{H}^1(\Xi_2) \to \mathbb{Z}[1/2]^2 \xrightarrow{\delta} \mathbb{Z}[1/4] \oplus \mathbb{Z}[1/2]^2 \to \check{H}^2(\Xi_2) \to \mathbb{Z}_3 \to 0. \tag{6.23}$$

The map $\delta : \check{H}^1(\Xi_1) \to \check{H}^2(\Xi_2, \Xi_1)$ is zero (since the boundary of every tile has a net of zero horizontal and zero vertical edges), so $\check{H}^1(\Xi_2) = \mathbb{Z}[1/2]^2$ and $\check{H}^2(\Xi_2)$ is an extension of $\mathbb{Z}[1/4] \oplus \mathbb{Z}[1/2]^2$ by $\mathbb{Z}_3$.

Which extension is seen by considering the class that counts chair tiles regardless of orientation. Since every chair tile consists of three arrow tiles, this class is one third of the generator of $\mathbb{Z}[1/4]$, so we have $H^2(\Omega) = \frac{1}{3}\mathbb{Z}[1/4] \oplus \mathbb{Z}[1/2]^2$ and $H^1(\Omega) = \mathbb{Z}[1/2]^2$.

This was a long route to a simple answer, but it *had* to be that way. As we have already seen, the generator of the rotationally invariant part of $H^2(\Omega)$ cannot be expressed in term of uncollared arrow tiles! The factor of 3 has to do with the conversion from arrows to chairs, which requires information about the neighborhood of each tile. This collaring information is captured in the torsion contribution to $\check{H}^2(\Xi_1)$. In turn, the only way to get torsion in $\check{H}^2(\Xi_1)$ is to have nonzero contributions to $\check{H}^2(\Xi_1, \Xi_0)$ and $\check{H}^1(\Xi_0)$ that almost cancel.

6.5. The pinwheel tiling

Recall that there are three spaces associated with a chair tiling: Ω_1, which is the translational orbit closure of a single tiling, Ω_{rot}, which is the Euclidean orbit closure of a single tiling, and $\Omega_0 = \Omega_1/\mathbb{Z}_4 = \Omega_{rot}/S^1$. For tilings where each tile appears in an infinite number of orientations, Ω_1 does

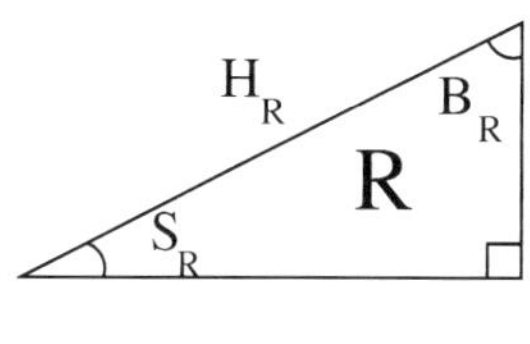
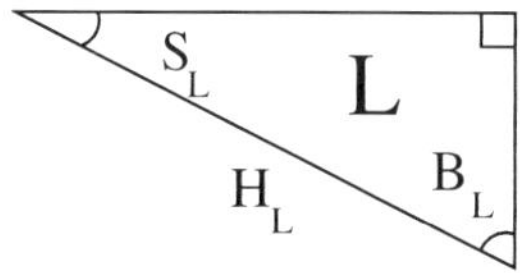

FIGURE 6.5. Right and left-handed pinwheel tiles with angles and hypotenuses marked

not make sense, but Ω_{rot} and Ω_0 do. As we saw in Chapter 4, the rational cohomology of Ω_{rot} is that of $\Omega_0 \times S^1$, but the integer cohomology may have torsion contributions to $\check{H}^2$ that come from tilings with rotational symmetry.

The Barge-Diamond construction is easily adapted to compute $\check{H}^*(\Omega_0)$. The Barge-Diamond complex is obtained from a single tiling by identifying points whose neighborhoods of size ϵ agree up to Euclidean motion. We have one tile cell for each tile type, up to rotation, an edge flap for each possible collared edge, up to rotation, and a vertex polygon for each possible vertex configuration, up to rotation. The inverse limit of this complex under substitution describes a tiling, up to rotation about the origin, and is therefore isomorphic to Ω_0. Some cells describe configurations with rotational symmetry, in which case points that are related by that symmetry are identified.

As before, we let S_0 be the union of the vertex polygons, S_1 be the union of the vertex polygons and edge disks, and S_2 the entire complex, and we take a homotopy $\bar{\sigma}$ of the substitution that sends S_0 to S_0 and S_1 to S_1, so we can consider the inverse limits $\Xi_0 = \varprojlim(S_0, \bar{\sigma})$, $\Xi_1 = \varprojlim(S_1, \bar{\sigma})$, and $\Xi_2 = \varprojlim(S_2, \bar{\sigma})$.

We work this computation for the pinwheel tiling. Before the invention of Barge-Diamond collaring, the pinwheel tiling space had defied calculation. There are 104 collared tiles up to rotation, making the Anderson-Putnam complex utterly unwieldy. (I tried repeatedly to analyze it, using all the tricks I knew, and never succeeded.) In particular, partial collaring gives only a slight simplification. Yet Barge-Diamond collaring works.

We call a pinwheel triangle with vertices at $(0,0)$, $(2,0)$ and $(2,1)$ *right-handed*, and one with vertices at $(0,0)$, $(2,0)$ and $(2,-1)$ *left-handed*. See Figure 6.5. We call the acute-angled vertices B_R, S_R, B_L, and S_L (for big-right, small-right, big-left, and small-left). We denote the two possible hypotenuses as H_R and H_L. Our complex has two tile cells, f_R and f_L, many edge flaps, and many, many vertex polygons. Fortunately, we only need to consider eventual ranges.

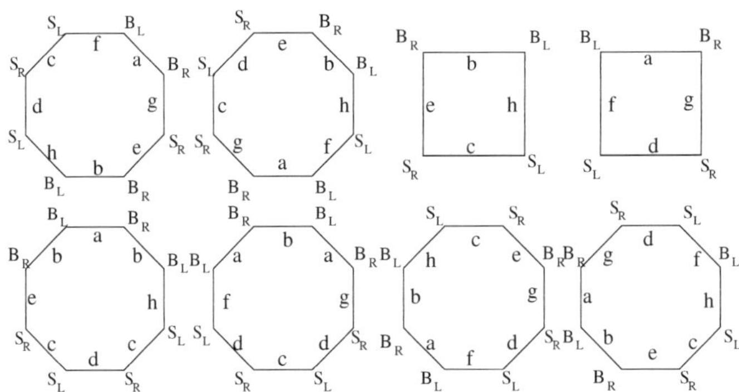

FIGURE 6.6. The eight pinwheel vertex polygons in $(S_0)_{\mathrm{ER}}$

Under substitution, right angles get split into smaller angles, so every vertex in the eventual range corresponds to a sequence of B_R, S_R, B_L and S_L. Exactly eight such patterns occur:
$$B_R B_L S_L S_R S_L B_L B_R S_R, \qquad B_L B_R S_R S_L S_R B_R B_L S_L,$$
$$B_L B_R S_R S_L B_L B_R S_R S_L, \qquad B_R B_L S_L S_R B_R B_L S_L S_R,$$
$$B_L B_R B_L B_R S_R S_L S_R S_L, \qquad B_R B_L B_R B_L S_L S_R S_L S_R,$$
$$B_R S_R S_L B_L B_R B_L S_L S_R, \qquad B_L S_L S_R B_R B_L B_R S_R S_L,$$
where we list the faces counterclockwise around the vertex. Under substitution, the first pattern becomes the second (and vice-versa), the third swaps with the fourth, the fifth swaps with the sixth, and the seventh swaps with the eighth. We represent all but the third and fourth as octagons, meeting the tile cells at points and the edge flaps along intervals. The third and fourth patterns have 180 degree rotational symmetry, so the quotients of these cells are quadrilaterals, rather than octagons, with patterns $B_L B_R S_R S_L$ and $B_R B_L S_L S_R$, respectively. See Figure 6.6.

After identifications, there are eight edges in $(S_0)_{\mathrm{ER}}$, corresponding to the transitions $B_R B_L$, $B_L B_R$, $S_R S_L$, $S_L S_R$, $B_R S_R$, $B_L S_L$, $S_R B_R$ and $S_L B_L$, and there are four vertices, namely B_R, B_L, S_R and S_L. The boundary maps are easily computed:

$$\partial_1 = \begin{pmatrix} -1 & 1 & 0 & 0 & -1 & 0 & 1 & 0 \\ 1 & -1 & 0 & 0 & 0 & -1^{\backprime} & 0 & 1 \\ 0 & 0 & -1 & 1 & 1 & 0 & -1 & 0 \\ 0 & 0 & 1 & -1 & 0 & 1 & 0 & -1 \end{pmatrix}$$

$$\partial_2 = \begin{pmatrix} 1 & 1 & 0 & 1 & 1 & 2 & 1 & 1 \\ 1 & 1 & 1 & 0 & 2 & 1 & 1 & 1 \\ 1 & 1 & 1 & 0 & 2 & 1 & 1 & 1 \\ 1 & 1 & 0 & 1 & 1 & 2 & 1 & 1 \\ 1 & 1 & 1 & 0 & 1 & 0 & 1 & 1 \\ 1 & 1 & 0 & 1 & 0 & 1 & 1 & 1 \\ 1 & 1 & 0 & 1 & 0 & 1 & 1 & 1 \\ 1 & 1 & 1 & 0 & 1 & 0 & 1 & 1 \end{pmatrix} \tag{6.24}$$

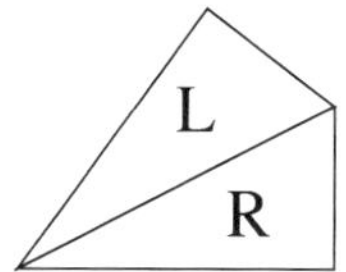

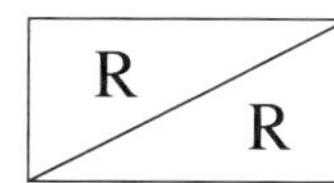

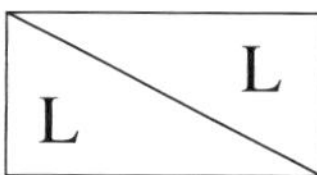

FIGURE 6.7. Three kinds of hypotenuses. The edge flap e_{RL} is a neighborhood of the hypotenuse in the first picture, while e_{RR} and e_{LL} are quotients of neighborhoods of the second and third hypotenuses by a 180 degree rotation.

and we compute $\check{H}^2(\Xi_0) = H^2((S_0)_{\mathrm{ER}}) = \mathbb{Z}^5$, $\check{H}^1(\Xi_0) = H^1((S_0)_{\mathrm{ER}}) = \mathbb{Z}^2$, $\check{H}^0(\Xi_0) = H^0((S_0)_{\mathrm{ER}}) = \mathbb{Z}$.

Edges in the pinwheel tiling come in two types: hypotenuses of length $\sqrt{5}$ and other edges of integer length. Substitution interchanges the two classes, so the contribution of each class to $\check{H}^*(\Xi_1, \Xi_0)$ must be the same, and we need only study the hypotenuses.

There are three kinds of hypotenuse edges, as shown in Figure 6.7: those where a right-handed tile meets a left-handed tile, those where two right-handed tiles meet, and those where two left-handed tiles meet. Call these e_{RL}, e_{RR}, and e_{LL}, respectively. Note that with e_{RR} and e_{LL} we must take the quotient by the 180 degree rotational symmetry, so the boundary of e_{RR} is just one hypotenuse of a right handed tile, not two. Our boundaries are $\partial(e_{RL}) = H_R + H_L$, $\partial(e_{RR}) = H_R$, and $\partial(e_{LL}) = H_L$, so $H^2 = \mathbb{Z}$ and $H^1 = 0$. Adding the contributions of the other class of edges, we get $H^2((S_1)_{\mathrm{ER}}, S_0) = \mathbb{Z}^2$. Furthermore, squared substitution multiplies these entries by 3, so $\check{H}^2(\Xi_1, \Xi_0) = \mathbb{Z}[1/3]^2$.

EXERCISE 6.14. *Squared substitution sends e_{RL}, e_{RR}, and e_{LL} to combinations of themselves. Find the matrix of this map. Likewise, squared substitution sends H_R and H_L to linear combinations of themselves. How do the eigenvalues of the two matrices compare? The (co)boundary map relates the eigenspaces with the same eigenvalue. The cohomology comes from the eigenspace of the first matrix that has no counterpart in the second.*

In the cohomology long exact sequence of (Ξ_1, Ξ_0), the coboundary map from $\check{H}^1(\Xi_0) \to \check{H}^2(\Xi_1, \Xi_0)$ is zero, so $\check{H}^2(\Xi_1) = \mathbb{Z}^5 \oplus \mathbb{Z}[1/3]^2$, $\check{H}^1(\Xi_1) = \mathbb{Z}^2$, and $\check{H}^0(\Xi_1) = 0$.

There are two tile cells, so S_2/S_1 is the wedge of two spheres. Thus $H^2(S_2, S_1) = \mathbb{Z}^2$ and $H^1(S_2, S_1) = H^0(S_2, S_1) = 0$. Under substitution, H^2 transforms by the transpose of the substitution matrix $M = \left(\begin{smallmatrix} 2 & 3 \\ 3 & 2 \end{smallmatrix}\right)$.

In the long exact sequence of the pair (Ξ_2, Ξ_1), the coboundary map $\delta : \check{H}^1(\Xi_1) \to H^2(\Xi_2, \Xi_1)$ is not trivial. Its image is all multiples of $(2, -2)$ (which is preserved by the substitution matrix), and the cokernel is $\mathbb{Z}[1/5] \oplus \mathbb{Z}_2$. This implies that $\check{H}^1(\Omega_0) = \mathbb{Z}$ and $\check{H}^2(\Omega_0) = \mathbb{Z}[1/5] \oplus \mathbb{Z}[1/3]^2 \oplus \mathbb{Z}^5 \oplus \mathbb{Z}_2$.

To get the rational cohomology of Ω_{rot}, we would need to identify the exceptional fibers of the fibration $\Omega_{rot} \to \Omega_0$. These come from the six pinwheel tilings with 180 degree rotational symmetry about the origin. Two correspond to the third and fourth vertex disks. Two correspond to the B and C hypotenuse edge flaps. Two are obtained from these edge flaps by substitution. These six configurations yield six exceptional fibers of the fibration $\Omega_{rot} \to \Omega_0$, hence a $\mathbb{Z}_2^5$ contribution to $H^2(\Omega_{rot})$. However, the exact computation of $\check{H}^2(\Omega_{rot})$ is complicated by the existence of torsion in $\check{H}^2(\Omega_0)$.

CHAPTER 7

Relaxing the rules II: Tilings without finite local complexity

Up to now, we have required our tiles to be polygons that meet full-edge to full-edge. This makes tilings rigid, in that a small motion to any one tile forces all of the neighboring tiles to undergo the same motion. In this chapter, we consider tilings in which the tiles are allowed to slide past one another.

7.1. Fault lines

The first substantial work on non-FLC tilings was done by Kenyon [**Ken**], who considered a substitution that was combinatorially equivalent to $a \to \left(\begin{smallmatrix} a & a & a \\ a & a & a \\ a & a & a \end{smallmatrix}\right)$, but in which one of the columns is shifted by an irrational distance relative to the other two*. Upon further substitution, the shift between the columns grows to an arbitrarily large multiple of the original irrational distance. In the limit, an infinite line of tile edges appears, along which tiles can face one another in an infinite variety of ways.

In geology, a *fault line* is the site of a past earthquake in which the ground shifted, so that the rocks on one side of the fault are not lined up with the rocks on the other side. Fault lines are also places where future shifts are possible, and even likely. In tiling theory, a fault line is a line (or line segment) where the tiles on one side appear shifted relative to the tiles on the other side, and where an appropriate shift of one side relative to the other gives another local pattern that is possible. We will see that this is an inevitable result of certain substitutions.

The first example with polygonal tiles was a generalization of the pinwheel tiling [**Sa1**]. Instead of using a 1–2–$\sqrt{5}$ right triangle, we use two sizes of a right triangle with irrational side lengths, each of which can be right- or left-handed. Substitution takes a small triangle to a big triangle, and a big triangle to a big triangle and four small triangles, as in Figure 7.1. Note that the two small triangles that share a hypotenuse are related by rotation, not reflection. On further substitution, what happens on the two sides of that hypotenuse do not have to match, and indeed they do not. Figure 7.2 shows a piece of that hypotenuse after several substitutions. It is plainly a fault line.

---

*The tile itself is not a polygon. Rather, it has straight edges on the left and right, while the top and bottom has fractal pieces that looked like the devil's staircase.

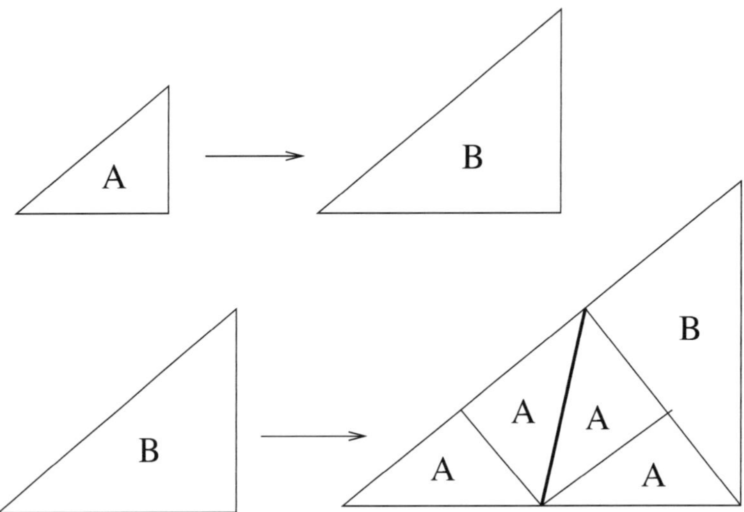

FIGURE 7.1. The substitution rule for the generalized pinwheel. On further substitution, a fault line will develop along the bold hypotenuse.

EXERCISE 7.1. *Deduce the lengths of the sides of the right triangles, assuming that the short side of the smaller triangle has length 1.*

Interestingly, this example meets the conditions of Goodman-Strauss' matching rules theorem [**GS**]. To wit: there exists a finite collection of tiles and a finite set of local matching rules such that these tiles tile the plane, but only in a manner that is locally equivalent to the generalized pinwheel. That is, a finite set of local rules forces a global hierarchical structure that in turn forces infinite local complexity. Truly bizarre!

Danzer [**Da**] extended the theory of non-FLC tilings and provided additional examples. Frank and Robinson [**Fr, FR**] have considered a large family of "direct product variation" (DPV) tilings. These are obtained from products of 1-dimensional substitutions by rearranging the positions of the tiles within an order-1 supertile. See Figure 7.3.

In all cases, the essential feature is the presence of fault lines. Along a fault line, evolution is described by two 1-dimensional substitutions. One describes what happens along one side of the fault line, while the other describes what happens along the other side.

Three Topologies. There are three metrics, and hence three topologies, that are frequently applied to tilings and tiling spaces.

The first metric was described in Chapter 1, and is applied to simple tilings. In it, two tilings are considered ϵ-close if they agree, up to a translation of size ϵ or less, on a ball of radius $1/\epsilon$ around the origin. In this topology, the closure of the (translational) orbit of a tiling is compact if and only if two conditions are met: (a) there are only a finite number of tile

FIGURE 7.2. Part of the generalized pinwheel

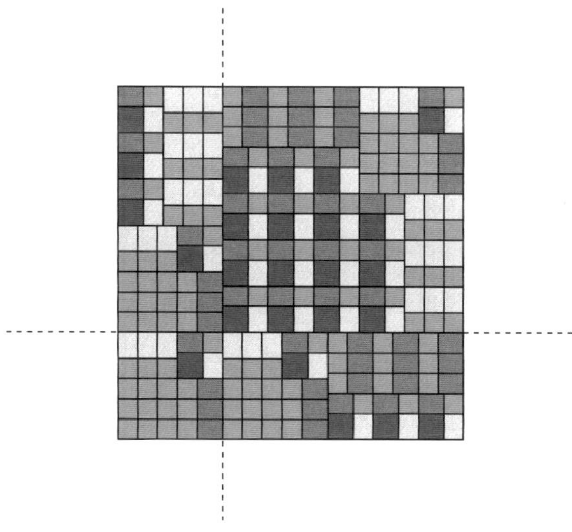

FIGURE 7.3. A patch of a DPV tiling with vertical and horizontal fault lines

types, up to translation, and (b) tiles can only meet in a finite number of ways, up to translation.

The second metric was described in Chapter 4. In it, one applies a metric to the group of rigid motions of the plane (e.g., defining an ϵ motion to be a translation of size $\leq \epsilon$ followed by a rotation by $\leq \epsilon$ about the origin). One then considers two tilings to be ϵ-close if they agree, up to an ϵ motion, on a ball of size $1/\epsilon$ about the origin. For tilings in which tiles appear in only finitely many orientations, this is equivalent to the first topology. However, it also handles pinwheel-like spaces appropriately. In this topology, the orbit closure of a tiling is compact if and only if the tiling has FLC.

Finally, one can use a metric in which two tilings are ϵ-close if they contain the same tiles out to distance $1/\epsilon$, and if each tile in the first tiling is within an ϵ-motion of the corresponding tile in the second tiling. Note that in this topology, an ϵ shear along a fault line yields a tiling that is ϵ-close to the original, while in the first two topologies it does not.

For tilings with FLC, the third topology is the same as the second, since it is impossible to apply a small rigid motion to one tile without applying the same motion to all of its neighbors. However, in the third topology the orbit closure of any tiling with finitely many tile types is *always* compact. To see sequential compactness, start with an arbitrary sequence of tilings and pick a subsequence in which the type, location and orientation of a tile near the origin converges. Then pick a subsequence of that subsequence in which the type, location and orientation of a second tile converges. Keep working outwards from the origin, and then apply a Cantor diagonalization argument to find a subsequence that converges everywhere.

EXERCISE 7.2. *Flesh out the details of this argument.*

In the third topology, FLC is not a topologically invariant property. See [**RS**] for a pair of tiling spaces, one with FLC and one without, that are topologically conjugate.

7.2. Analyzing a fault line — 1 dimensional dynamics

As mentioned before, every fault line is associated with a pair of one-dimensional substitutions describing what happens on the two sides of the fault. In this section we'll see how a pair of substitutions with the same stretching factor can generate a fault line. In the next section we will examine a simple 2-dimensional substitution whose associated one-dimensional substitutions are the ones of this section. Let

$$\sigma_1(a) = ba, \qquad \sigma_1(b) = aaa, \qquad \sigma_2(a) = ab, \qquad \sigma_2(b) = aaa. \qquad (7.1)$$

Both substitutions have substitution matrix $\left(\begin{smallmatrix} 1 & 3 \\ 1 & 0 \end{smallmatrix}\right)$, with dominant (Perron-Frobenius) eigenvalue $\lambda = (1 + \sqrt{13})/2 \approx 2.3028$, the larger root of the equation $\lambda^2 - \lambda - 3 = 0$. For a self-similar tiling, the a tile can be given length λ while the b tile has length 3. Note that for any word W, $\sigma_2(W)$ is a cyclic permutation of $\sigma_1(W)$, obtained by removing an a from the end and sticking it on the beginning. If W is a bi-infinite word, then $\sigma_1(W)$ and $\sigma_2(W)$ are the same, up to translation by the length of a. This implies that Ω_{σ_1} and Ω_{σ_2} are the same space, which we denote Ω_σ.

Suppose we have a tiling with rectangular tiles a and b of widths λ and 3, respectively, and suppose that σ_1 acts as a substitution on lower edges and σ_2 acts as a substitution on upper edges. At a horizontal fault line, the evolution above the line is governed by σ_1 and the evolution below the line is governed by σ_2. If at some stage there is a pair of exactly aligned a tiles, one above the line and one below, then on successive substitutions we will see $\left(\begin{smallmatrix} ba \\ ab \end{smallmatrix}\right)$, $\left(\begin{smallmatrix} aaaba \\ abaaa \end{smallmatrix}\right)$, $\left(\begin{smallmatrix} bababaaaaba \\ abaaaababab \end{smallmatrix}\right)$, $\left(\begin{smallmatrix} aaabaaaabaaaababababaaaaba \\ abaaaababababaaaaabaaaabaaa \end{smallmatrix}\right)$. Note that in the first and third substitution, the a tiles are found more on the right of the top row and on the left of the bottom row, while in the second and fourth substitutions they are found more on the left of the top row and the right of the bottom row. The reason is that the difference between the number of a tiles up to a certain point grows like the second eigenvalue of the substitution matrix, namely $1 - \lambda \approx -1.3$. As we continue to iterate, this discrepancy grows without bound.

Now pick a point p along the fault line. If there are m more a tiles in the top row than the bottom up to p, then the left edges of the tiles on the top row will be offset by λm (mod 3) relative to the left edges of the tiles on the bottom row. Since the discrepancy can only change in steps of 1 as we move along the line, it takes on all integer values between 0 and m as we move from the left edge of the pattern to p. Since m is unbounded and λ is irrational, this means that the possible offsets of tiles in the top and bottom rows takes on a dense set of values in the limit of infinite substitution. In fact the left endpoints of upper a tiles are dense in the lower b tiles, because the only way for the discrepancy to grow from m to $m+1$ is for an additional a tile to appear along the top with a b tile below it. Thus every possible adjacency between an upper a an a lower b can occur in the orbit closure. By applying the substitution once more we get every possible adjacency between any upper and lower tiles.

Note how the formation of the fault line depends on the second eigenvalue of the substitution matrix. If the second eigenvalue were less than one, then the discrepancy in the number of any species of tile would be bounded, and the offsets between tiles would take on only a finite number of values. As a result, the FLC condition would be preserved.

A substitution is said to be *Pisot* if all but one of the eigenvalues of the substitution matrix have magnitudes strictly between 0 and 1. If the two substitutions associated with a fault line are Pisot, then the FLC property is preserved [**FR**]; tiles may not meet full-edge to full-edge, but they only meet in a finite number of ways. However, if the substitutions are not Pisot, then differences in distributions of lengths on opposite sides of the fault line will generally grow as λ_2^n, where λ_2 is the second largest eigenvalue and n is the number of substitutions, and the FLC property can be lost.

7.3. A simple 2-dimensional example

Consider a 2-dimensional tiling with two rectangular tiles. Both the A and B tiles have height 1, but the A tile has width $\lambda = (1 + \sqrt{13})/2$ and the B tile has width 3. We consider the self-affine[†] substitution Σ shown in Figure 7.4.

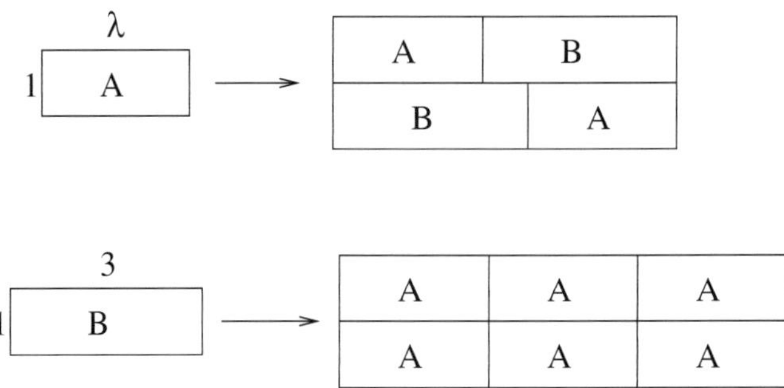

FIGURE 7.4. A substitution with a horizontal fault line

A tiling is *allowed* by the substitution if each of its finite patches of tiles can be found in some n-supertile. The smallest closed set (under the third topology) containing all allowed tilings is called the *substitution tiling space* Ω_Σ. We will see that measure theoretically, almost all of the tilings in Ω_Σ are allowed by the substitution, but that the ones that are not are the most interesting!

Whenever two supertiles meet along a horizontal boundary, applying Σ changes the bottom of the top supertile by σ_1 and the top of the bottom supertile by σ_2. By the results of the last section, the tilings in Ω_Σ do not have FLC.

Not every row is subject to arbitrary shears. The rows themselves are labeled by infinite sequences of binary digits, describing their hierarchy in the vertical substitution. The label in the nth spot is 0 if the row is in the lower $(n-1)$-supertile of its n-supertile and 1 if it is in the upper $(n-1)$-supertile. If the labels of two adjacent rows differ only in the first digit, then the label of the upper row begins with a 1 and the label of the lower row begins with a 0. One can see that the sequence of tiles in the two rows are identical, with the upper row offset horizontally by λ. If they differ only in the first two digits, then the dyadic label of the upper row begins with 01 and the label of the lower begins with 10, and the upper row is the same as the lower row, but offset by $\lambda^2 - \lambda$. If they differ only in the first three digits, the upper and lower labels begin with 001 and 110 respectively, and then the rows are offset by $\lambda^3 - \lambda^2 - \lambda$. If they differ in the first n digits (and agree thereafter), then the upper and lower labels begin with $0^{n-1}1$ and $1^{n-1}0$ resp., and they are offset by $\lambda^n - \lambda^{n-1} - \cdots - \lambda$. (In general, one sees the new offset as λ times the previous offset, minus λ.) However, in

---

[†]So far we have considered *self-similar* substitutions, in which tiles are stretched by a constant factor and then subdivided. In a *self-affine* substitution, we apply an expansive linear transformation and then subdivide. In this case the linear transformation stretches horizontally by λ and vertically by 2.

some tilings there exists a row with label $1111\ldots$ and an adjacent row above it with label $0000\ldots$. These rows do not have to have the same sequence of tiles, and their offset is arbitrary.

Put another way, all tilings in the tiling space contain horizontal fault lines separating identical rows of tiles, offset by arbitrarily large amounts. However, in a some tilings there exists a fault line with infinite shear, meaning that the tiling above the fault is unrelated to the tiling below the fault. We call this a *critical fault line*.

Tilings with critical fault lines have measure zero with respect to all translation-invariant measures, and hence have no effect on measure theoretic properties of the tiling space. To see this, consider the probability that the critical fault line has y coordinate between 0 and 1. By translation invariance, this probability must be the same as that of having the line between 1 and 2, or 2 and 3, etc., and so must equal zero. The probability of having a critical fault line *somewhere* is the sum of the probability of having it between 0 and 1, 1 and 2, 2 and 3, -1 and 0, etc., namely $0 + 0 + \cdots = 0$.

7.4. The local topology of Ω_Σ

In the chair tiling space, the neighborhood of any point is a 2-disk times a Cantor set. In the pinwheel tiling space, the neighborhood of any point is a 3-disk times a Cantor set. What does the neighborhood of a point $T \in \Omega_\Sigma$ look like?

The answer depends on whether T is a tiling with a critical fault line. If it is, then a neighborhood of T looks like a 3-disk times a Cantor set. We have three continuous degrees of freedom, from moving the tiling up or down, moving the lower half of the tiling left or right, and moving the upper half of the tiling left or right. We also have infinitely many discrete choices to make for the sequences of tiles above and below the critical fault line. Put another way, a neigborhood in Ω_Σ looks like a neighborhood in $\mathbb{R}$ (describing the height of the critical fault line) times the product of two neighborhoods in the 1-dimensional tiling space Ω_σ (describing the rows above and below the critical fault).

If T does not have a critical fault line, then most of a small neighborhood of T looks like a 2-disk crossed with a Cantor set. We can move the tiling side-to-side or up-and-down, and we can make discrete choices about vertical hierarchy far from the origin, and about the sequence of tiles in the far left or far right of any one row. In other words, we have a neighborhood in the dyadic solenoid $\varprojlim(S^1, \times 2)$, describing the vertical hierarchy, times a neighborhood in Ω_σ.

But that's not everything. In the ϵ-neighborhood of T there are tilings with critical fault lines located more than $1/\epsilon$ from the origin. If T' is such a tiling at a distance $\delta < \epsilon$ from T, then the ϵ-neighborhood of T contains the $(\epsilon - \delta)$-neighborhood of T', and therefore contains a piece that looks a 3-disk times a Cantor set.

In summary, Ω_Σ is an odd mixture of 2-dimensional and 3-dimensional pieces. The 3-dimensional pieces have measure zero, but they are dense, and you can't understand the topology of Ω_Σ without confronting these pieces.

7.5. Inverse limit structures

Despite having pieces of different dimensions, Ω_Σ can be expressed the the inverse limit of a CW complex. The first construction [**FS**] exhibited Ω_Σ as the inverse limit of a branched 4-manifold S_4, under a map that was not onto. In this section we use Barge-Diamond collaring to exhibit Ω_Σ as the inverse limit of a CW complex with both 2-dimensional and 3-dimensional pieces. We will then use this inverse limit structure to compute the cohomology of Ω_Σ. Before doing so, we need to introduce a new topological structure.

Let X and Y be topological spaces. The *join* of X and Y, denoted $X * Y$, is the space $X \times Y \times [0,1]/ \sim$, where $(x, y, 0) \sim (x, y', 0)$ and $(x, y, 1) \sim (x', y, 1)$. If we let t denote the coordinate in $[0, 1]$, we are collapsing Y to a point at $t = 0$ and collapsing X to a point at $t = 1$.

THEOREM 7.1. *If X and Y are finite, connected, 1-dimensional CW complexes, then $H^0(X * Y) = \mathbb{Z}$, $H^1(X * Y) = H^2(X * Y) = 0$ and $H^3(X * Y) = H^1(X) \otimes H^1(Y)$.*

PROOF. We use the Mayer-Vietoris sequence [**Hat**] in reduced cohomology

$$\to \tilde{H}^k(U \cup V) \to \tilde{H}^k(U) \oplus \tilde{H}^k(V) \to \tilde{H}^k(U \cap V) \to \tilde{H}^{k+1}(U \cup V) \to, \quad (7.2)$$

with $U = \{(x, y, t) | t < \frac{2}{3}\}$ and $V = \{(x, y, t) | t > \frac{1}{3}\}$. U retracts to $X \times y_0 \times \{0\}$, where y_0 is an arbitrary point in Y. Likewise, V retracts to $x_0 \times Y \times \{1\}$ and $U \cap V$ retracts to $X \times Y \times \{\frac{1}{2}\}$, so the Mayer-Vietoris sequence (7.2) reduces to

$$\to \tilde{H}^k(X * Y) \to \tilde{H}^k(X) \oplus \tilde{H}^k(Y) \to \tilde{H}^k(X \times Y) \to \tilde{H}^{k+1}(X * Y) \to. \quad (7.3)$$

Since the cohomology of X is finitely generated and free, the Künneth formula [**Hat**] says that $H^*(X \times Y) = H^*(X) \otimes H^*(Y)$, or more precisely that

$$\begin{aligned}
H^0(X \times Y) &= \mathbb{Z}, \\
H^1(X \times Y) &= H^1(X) \oplus H^1(Y), \\
H^2(X \times Y) &= H^1(X) \otimes H^1(Y).
\end{aligned} \quad (7.4)$$

Plugging that into (7.3) gives

$$\begin{aligned}
0 \to \tilde{H}^0(X * Y) &\to 0 \to 0 \\
\hookrightarrow H^1(X * Y) &\to H^1(X) \oplus H^1(Y) \to H^1(X) \oplus H^1(Y) \\
\hookrightarrow H^2(X * Y) &\to 0 \to H^1(X) \otimes H^1(Y) \\
\hookrightarrow H^3(X * Y) &\to 0 \to 0.
\end{aligned} \quad (7.5)$$

This immediately gives $H^3(X * Y) = H^1(X) \otimes H^1(Y)$ and $\tilde{H}^0(X * Y) = 0$. The map from $H^1(X) \oplus H^1(Y)$ to itself is the identity, so $H^1(X * Y) = H^2(X * Y) = 0$. $\qquad\square$

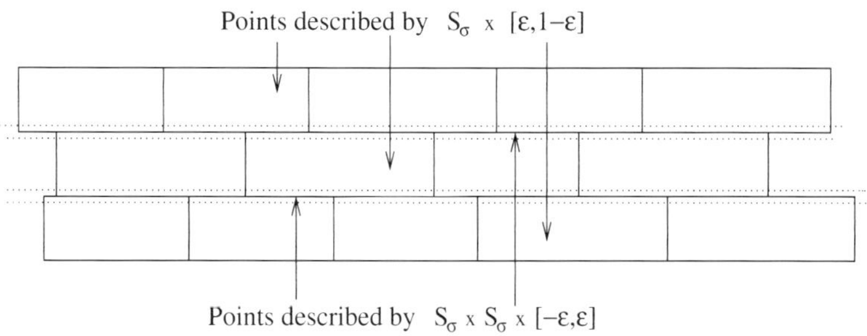

FIGURE 7.5. Describing points in a tiling without FLC.

Now we construct the Barge-Diamond complex for Ω_Σ. Pick an $\epsilon < \frac{1}{4}$. Let S_σ denote the Barge-Diamond complex for Ω_σ, with collaring out to distance ϵ. We want a space S_Σ that describes neighborhoods of points in possible tilings out to a horizontal distance ϵ and a vertical distance ϵ.

This consists of two pieces. Points that are at least a distance ϵ from a horizontal edge are described by $S_\sigma \times [\epsilon, 1 - \epsilon]$. The first coordinate describes where they sit left-to-right within a tile, and possibly describes their neighbor to the right or left, while the second coordinate gives their vertical position in a tile. Points that are within ϵ of a horizontal edge are described by $S_\sigma \times S_\sigma \times [-\epsilon, \epsilon]$, where the first coordinate describes the tiling (out to horizontal distance ϵ) just below the edge, the second coordinate describe the tiling just above the edge, and the third coordinate gives the height of our reference point, relative to the edge. On the boundary of these two regions, $(x, y, \epsilon) \in S_\sigma \times S_\sigma \times [-\epsilon, \epsilon]$ is identified with $(y, \epsilon) \in S_\sigma \times [\epsilon, 1-\epsilon]$ and $(x, 1 - \epsilon)$ is identified with $(x, y, -\epsilon)$.

Note that (x, y, ϵ) is identified with (x', y, ϵ) and that $(x, y, -\epsilon)$ is identified with $(x, y', -\epsilon)$. The image of $S_\sigma \times S_\sigma \times [-\epsilon, \epsilon]$ under these identifications is $S_\sigma * S_\sigma$, only with the interval $[-\epsilon, \epsilon]$ instead of $[0, 1]$. We call this image S_1.

THEOREM 7.2. *The nontrivial cohomology groups of Ω_Σ are as follows:*

$$
\begin{aligned}
\check{H}^0(\Omega_\Sigma) &= 0 \\
\check{H}^1(\Omega_\Sigma) &= \mathbb{Z}[1/2] \\
\check{H}^2(\Omega_\Sigma) &= H^1(\Omega_\sigma) \otimes \mathbb{Z}[1/2] \\
\check{H}^3(\Omega_\Sigma) &= H^1(\Omega_\sigma) \otimes H^1(\Omega_\sigma),
\end{aligned}
\tag{7.6}
$$

where $H^1(\Omega_\sigma)$ is the direct limit of $\mathbb{Z}^2$ under the matrix $\left(\begin{smallmatrix} 1 & 1 \\ 3 & 0 \end{smallmatrix}\right)$.

PROOF. The cohomology of Ω_σ is given by Corollary 6.6, since the substitution matrix of σ is $\left(\begin{smallmatrix} 1 & 3 \\ 1 & 0 \end{smallmatrix}\right)$.

Ω_Σ is the inverse limit of S_Σ by the substitution map. We will compute $H^*(S_\Sigma)$ and then take the direct limit of this cohomology by a map $\bar\Sigma$ that is homotopic to the substitution map and that takes S_1 to itself.

The relative cohomology groups $H^*(S_\Sigma, S_1)$ are easily computed. The quotient space S_Σ/S_1 is $S_\sigma \times [\epsilon, 1-\epsilon]$, with all points (y, ϵ) and $(x, 1-\epsilon)$ identified. It is not hard to see that $H^1(S_\Sigma, S_1) = H^1(S^1) = \mathbb{Z}$ and that $H^2(S_\Sigma, S_1) = H^1(S_\sigma)$.

EXERCISE 7.3. *Compute $H^*(S_\Sigma/S_1)$ using a Mayer-Vietoris sequence with $U = \{(x,t)|\frac{1}{4} < t < \frac{3}{4}\}$ and $V = \{(x,t)|t < \frac{1}{3} \text{ or } t > \frac{2}{3}\}$. Note that U retracts to S_σ, V retracts to a point, and $U \cap V$ retracts to two copies of S_σ.*

The cohomology of S_1 is given by Theorem 7.1. The cohomology exact sequence of the pair (S_Σ, S_1) then reads

$$
\begin{array}{ccccc}
0 & \to & \tilde{H}^0(S_\Sigma) & \to & 0 \\
\mathbb{Z} & \to & H^1(S_\Sigma) & \to & 0 \\
H^1(S_\sigma) & \to & H^2(S_\Sigma) & \to & 0 \\
0 & \to & H^3(S_\Sigma) & \to & H^1(S_\sigma) \otimes H^1(S_\sigma) \to 0.
\end{array}
\qquad (7.7)
$$

This gives $H^0(S_\Sigma) = \mathbb{Z}$, $H^1(S_\Sigma) = \mathbb{Z}$, $H^2(S_\Sigma) = H^1(S_\sigma)$ and $H^3(S_\Sigma) = H^1(S_\sigma) \otimes H^1(S_\sigma)$. The generator of H^0 counts points, the generator of H^1 counts displacement from one horizontal edge to the next, the generators of H^2 count uncollared tiles of the two types, and the generators of H^3 are products of terms that count tiles above and below a fault line.

Under substitution, the generator of H^1 gets multiplied by the vertical stretching factor of 2, so the limit is $\mathbb{Z}[1/2]$. H^2 transforms by the substitution matrix of Σ which is twice the substitution matrix of σ, so the limit is $H^1(\Omega_\sigma) \otimes \mathbb{Z}[1/2]$. Substitution acts on $S_\sigma \times S_\sigma$ by $\sigma_1 \times \sigma_2$. Since the direct limit of $H^1(S_\sigma)$ by either σ_1^* or σ_2^* is $H^1(\Omega_\sigma)$, the direct limit of $H^1(S_\sigma) \otimes H^1(S_\sigma)$ is $H^1(\Omega_\sigma) \otimes H^1(\Omega_\sigma)$. $\square$

Theorem 7.2 vividly shows the cohomological signature of a critical fault line. Most of Ω_Σ is 2-dimensional and looks like the product of Ω_σ and the dyadic solenoid $\varprojlim(S^1, \times 2)$. Since the dyadic solenoid has $\check{H}^1 = \mathbb{Z}[1/2]$, the product of Ω_σ and the dyadic solenoid has $\check{H}^1 = H^1(\Omega_\sigma) \oplus \mathbb{Z}[1/2]$, $\check{H}^2 = H^1(\Omega_\sigma) \otimes \mathbb{Z}[1/2]$ and $\check{H}^3 = 0$. The 3-dimensional part of Ω_Σ, describing tiles with critical fault lines, gives a $\check{H}^1(\Omega_\sigma) \otimes \check{H}^1(\Omega_\sigma)$ contribution to $\check{H}^3$, but also kills the $\check{H}^1(\Omega_\sigma)$ term in $\check{H}^1$.

7.6. A more involved example

We next consider a substitution that admits two different types of critical fault lines (but not in the same tiling), and see how this affects the cohomology. Let σ_1, σ_2, and Ω_σ be as before. Our last example was almost the product of Ω_σ with the dyadic solenoid; this example will almost be the product of Ω_σ and Ω_ρ, where ρ is the period-doubling substitution on two letters: $\rho(x) = xX$, $\rho(X) = xx$.

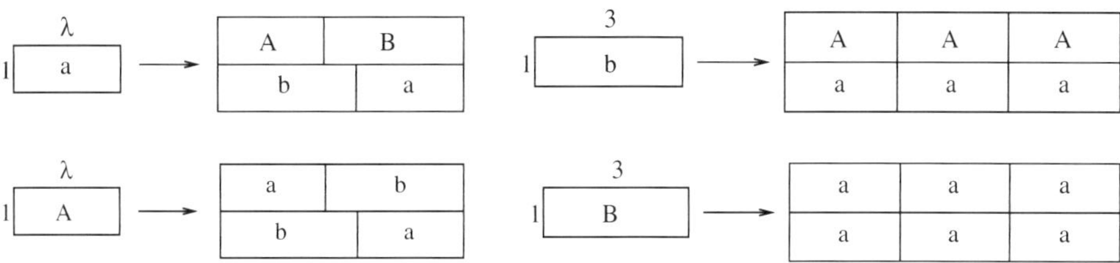

FIGURE 7.6. A substitution with two kinds of horizontal fault lines

EXERCISE 7.4. *Contruct the Barge-Diamond complex for ρ. Call this complex S_ρ, and call the sub-complex of vertex flaps $S_{\rho,0}$. Compute $H^*(S_{\rho,0})$.*

EXERCISE 7.5. *Compute $H^1(\Omega_\rho)$.*

Our example has four tiles: a, A, b and B. All tiles are rectangles of height 1. The a and A tiles have width λ, and the b and B tiles have width 3. The capital letters should be viewed as products of the a and b tiles in Ω_σ with the X tile in Ω_ρ, and the lower case letters are products with x. The 2-dimensional substitution (which we call Σ) is shown in Figure 7.6. Note that the bottom of each supertile is made of lower case tiles, while the top may be either capital or lower case. If a tiling has a critical fault line, then the row above the critical fault is made of lower case tiles, but the row below can be either capital or lower case. Put another way, there are only two vertex flaps in the eventual range of $S_{\rho,0}$: v_{Xx} and v_{xx}. Each corresponds to a possible type of critical fault line.

We construct the Barge-Diamond complex of Ω_Σ by collaring a distance ϵ vertically and ϵ horizontally. Let S_1 be the collared points within ϵ of a horizontal line, as in Figure 7.5. S_1 consists of three pieces, corresponding to the three vertex flaps in $S_{\rho,0}$.

The piece corresponding to v_{xx}, which we denote F_{xx}, is isomorphic to $S_\sigma * S_\sigma$. Likewise, F_{Xx} is isomorphic to $S_\sigma * S_\sigma$. F_{xX} is only 2-dimensional, since there are only a finite number of patterns where lower case tiles lie below capitalized tiles. Its structure is complicated, but can be ignored, since F_{xX} is not in the eventual range of S_1.

EXERCISE 7.6. *In $(S_1)_{ER}$, let U be a neighborhood of F_{xx} and let V be a neighborhood of F_{Xx}. Use the Mayer-Vietoris sequence (7.2) to show that $H^1((S_1)_{ER}) = 0$, $H^2((S_1)_{ER}) = H^1(S_\sigma)$, and $H^3((S_1)_{ER}) = [H^1(S_\sigma) \otimes H^1(S_\sigma)] \oplus [H^1(S_\sigma) \otimes H^1(S_\sigma)]$.*

To get the cohomology of $\Xi_1 = \varprojlim(S_1, \bar{\Sigma})$, we take the direct limit of $H^*((S_1)_{ER})$.

EXERCISE 7.7. *Show that substitution acts by σ_1^* (not σ_2^*) on $H^2((S_1)_{ER})$ and acts by $(\sigma_1^* \otimes \sigma_2^*) \oplus (\sigma_1^* \otimes \sigma_2^*)$ on $H^3((S_1)_{ER})$. Since $\varinjlim(H^1(S_\sigma), \sigma_1^*) = \varinjlim(H^1(S_\sigma), \sigma_2^*) = H^1(\Omega_\sigma)$, this implies that $H^2(\Xi_1) = \check{H}^1(\Omega_\sigma)$ and that $\check{H}^3(\Xi_1) = [H^1(\Omega_\sigma) \otimes H^1(\Omega_\sigma)] \oplus [H^1(\Omega_\sigma) \otimes H^1(\Omega_\sigma)]$.*

What's left is to compute $H^*(\Xi_\Sigma, \Xi_1)$ and to use the long exact sequence of the pair (Ξ_Σ, Ξ_1). The quotient space S_Σ/S_1 is two copies of $S_\sigma \times [\epsilon, 1-\epsilon]$ with all points with $t = \epsilon$ or $t = 1 - \epsilon$ identified. This has $H^1(S_\Sigma, S_1) = H^1(S_\Sigma/S_1) = \mathbb{Z}^2$, with generators counting loops in the t direction in the two pieces, and $H^2(S_\Sigma, S_1) = H^2(S_\Sigma/S_1) = H^1(S_\sigma) \oplus H^1(S_\sigma)$, again with one term from each piece.

Since the H^1 terms refer only to t, substitution on them looks just like ρ^* acting on $H^1(S_\rho, (S_\rho)_0)$. In other words, Σ^* acts by the transpose of the ρ-substitution matrix $M_\rho = \left(\begin{smallmatrix} 1 & 2 \\ 1 & 0 \end{smallmatrix}\right)$, and the direct limit is $\check{H}^1(\Xi_\Sigma, \Xi_1) = \varinjlim(\mathbb{Z}^2, M_\rho^T) = \check{H}^1(\Omega_\rho) = \mathbb{Z} \oplus \mathbb{Z}[1/2]$.

On $H^2(S_\Sigma, S_1)$, substitution acts by the matrix $\left(\begin{smallmatrix} \sigma_2^* & \sigma_1^* \\ \sigma_2^* + \sigma_1^* & 0 \end{smallmatrix}\right)$. The maps σ_1^* and σ_2^* on $H^1(S_\sigma)$ are actually the same (see exercise 7.8, below), so we have $\sigma^* \otimes M_\rho^T$, and the direct limit is $\check{H}^1(\Omega_\sigma) \otimes \check{H}^1(\Omega_\rho)$.

EXERCISE 7.8. *Explicitly compute the actions of σ_1^* and σ_2^* on $H^1(S_\sigma)$ and show that they are the same. This is related to the fact that σ_1 and σ_2, acting on Ω_σ, are actually homotopic.*

Finally we put it all together. The cohomology long exact sequence of the pair (Ξ_Σ, Ξ_1) reads

$$
\begin{aligned}
0 \to \mathbb{Z} \quad &\to \quad \check{H}^0(\Omega_\Sigma) \quad \to \quad && 0 \\
\check{H}^1(\Omega_\rho) \quad &\to \quad \check{H}^1(\Omega_\Sigma) \quad \to \quad && 0 \\
\check{H}^1(\Omega_\sigma) \otimes \check{H}^1(\Omega_\rho) \quad &\to \quad \check{H}^2(\Omega_\Sigma) \quad \to \quad && \check{H}^1(\Omega_\sigma) \\
0 \quad &\to \quad \check{H}^3(\Omega_\Sigma) \quad \to \quad && [\check{H}^1(\Omega_\sigma) \otimes H^1(\Omega_\sigma)]^2 \to 0.
\end{aligned}
\qquad (7.8)
$$

The first line determines $\check{H}^0(\Omega_\Sigma)$, the second line determines $\check{H}^1(\Omega_\Sigma)$, and the fourth line determines $\check{H}^3(\Omega_\Sigma)$. The only subtlety is the third line. This short exact sequence splits, because the maps only involve restrictions in the vertical direction, with the $\check{H}^1(\Omega_\sigma)$ factor coming along for the ride. The final result is

$$
\begin{aligned}
\check{H}^0(\Omega_\Sigma) &= \mathbb{Z} \\
\check{H}^1(\Omega_\Sigma) &= \check{H}^1(\Omega_\rho) = \mathbb{Z} \oplus \mathbb{Z}[1/2] \\
\check{H}^2(\Omega_\Sigma) &= \check{H}^1(\Omega_\sigma) \otimes [\check{H}^1(\Omega_\rho) \oplus \mathbb{Z}] \\
\check{H}^3(\Omega_\Sigma) &= [\check{H}^1(\Omega_\sigma) \otimes \check{H}^1(\Omega_\sigma)]^2.
\end{aligned}
\qquad (7.9)
$$

In our simple example, the presence of a critical fault line added an $\check{H}^1(\Omega_\sigma) \otimes \check{H}^1(\Omega_\sigma)$ term to $\check{H}^3(\Omega_\Sigma)$ and killed off the $\check{H}^1(\Omega_\sigma)$ term in $\check{H}^1(\Omega_\Sigma)$ that one might have expected. The second possible critical fault line adds another $\check{H}^1(\Omega_\sigma) \otimes \check{H}^1(\Omega_\sigma)$ term to $\check{H}^3(\Omega_\Sigma)$, but there is no $\check{H}^1(\Omega_\sigma)$ term left to kill off in $\check{H}^1$. Instead, the critical fault adds such a term to $\check{H}^2$. If we had an example with n possible critical fault lines, we should expect $\check{H}^3$ to equal $[\check{H}^1(\Omega_\sigma) \otimes \check{H}^1(\Omega_\sigma)]^n$, and for $\check{H}^2$ to equal $\check{H}^1(\Omega_\sigma) \otimes [\check{H}^1(\Omega_\rho) \oplus \mathbb{Z}^{n-1}]$. In [**FS**] it's proven that, under mild technical assumptions, this is indeed the case.

7.7. Conclusions

This does not exhaust the study of tilings without FLC. Far from it. We have not considered tilings with fault lines in multiple directions. We have not considered tilings in which the rows are fundamentally different. There may even be tilings with fault lines where FLC is lost, but not every possible offset of neighboring tiles is possible.

As of this writing, none of these tilings are fully understood. What *is* understood are the tools that can be brought to bear. Barge-Diamond collaring is key, as are relative cohomology, Mayer-Vietoris sequences, and inverse limits. I wish I could tell you how to combine these tools in new and interesting ways. Then again, maybe that's best left to others — perhaps you.

APPENDIX A

Solutions to selected exercises

Chapter 1.

1.1. Ω_T looks like a slinky. It has three path components, with two circles and an infinite helix that interpolates between them. One circular component is the all white tiling (and its translates), and the other is the all black tiling. This space embeds in the plane as the graph of $r = \frac{1+2e^\theta}{1+e^\theta}$, plus the circles $r = 1$ and $r = 2$.

1.2. With probability one, Ω_T consists of *all* tilings by black and white tiles, including the infinitely improbable "all white" tiling. Recall that $T' \in \Omega_T$ if every patch of T' is found somewhere in T. But for every patch P of an arbitrary tiling T', the probability of that patch occuring somewhere is one.

1.3. The population of $\sigma^n(a_j)$ is $M^n e_j$, so the length is of $\sigma^n(a_j)$ is $LM^n e_j = \lambda_{PF}^n L e_j = \lambda_{PF}^n L_j$.

1.4. Decompose $e_j = \sum c_j \mathbf{v}_j$ as a linear combination of right-eigenvectors of M, with $\mathbf{v}_1 = R$. Then $M^n e_j = \sum c_j \lambda_j^n \mathbf{v}_j$. The total number of tiles is $(1, 1, \ldots, 1) M^n e_j$. Since $c_1 \ne 0$ and $|\lambda_1| > |\lambda_j|$ for $j > 1$, $M^n e_j$ is dominated by the first term in the sum, and $M^n e_j / [(1, \ldots, 1) M^n e_j] \to \mathbf{v}_1 / (1, \ldots, 1)\mathbf{v}_1$, the i-th term of which is $R_i / \sum_k R_k$.

1.5. Since every supertile for σ^k is also a supertile for σ, the set of admissible words for σ^k is a subset of the set of admissible words for σ. Conversely, if a word is found in $\sigma^n(a_j)$, then it is also found in supertiles of all orders greater than n, and in particular in orders that are multiples of k.

1.6. Order 1: *b.ba.ab.ab.ba.ba.ab.ab.ba.ab.ba.ba.*
Order 2: *b.baab.abba.baab.abba.abba.ba.*
Order 3: *b.baababba.baababba.abbaba.*
Order 4: *bbaababba.baababbaabbaba.*
Order 5: You can't tell whether the two order-4 supertiles belong to the same order-5 supertile or not.

1.7. Since T contains a and b supertiles of arbitrarily high order, every σ-admissible patch is found somewhere in T. Since $b.a$ is itself a σ-supertile and every patch in T is part of $\sigma^n(b.a)$, every patch in T is σ-admissible. So the conditions for T' to be in Ω_T and in Ω_σ are the same.

109

1.8.   Use capital letters to denote Thue-Morse tiles and lower-case letters for period-doubling tiles. Let $(A)B$ denote a B tile that is preceded by an A tile, etc. Then $f \circ \sigma_{TM}((A)B) = f((AB)BA) = ab = \sigma_{PD}(b) = \sigma_{PD} \circ f((A)B)$, and $f \circ \sigma_{TM}((A)A) = f((AB)AB) = bb = \sigma_{PD}(a) = \sigma_{PD} \circ f((A)A)$. The situation for $(B)A$ and $(B)B$ is similar. This shows that f of a Thue-Morse supertile is a period-doubling supertile, hence that f of a Thue-Morse tiling is a valid period-doubling tiling. That f is a 2:1 cover is then easy. If T is a period-doubling tiling, then $f^{-1}(T)$ consists of two aligned tilings that disagree at *every* position.

Chapter 2.

2.1.   In 1 dimension, pick an n such that $\lambda^n L_{min} > 2L_{max}$, where L_{min} and L_{max} are the lengths of the smallest and largest tiles. Collaring a tile specifies a distance of at least L_{min} in each direction from the tile. Applying a substitution n times then specifies a distance at least $2L_{max}$ around the supertile, which therefore specifies all of the collared tiles that touch the supertile. The argument in higher dimensions uses the largest tile diameter in place of L_{max}, but defining the right quantity to replace L_{min} is slightly trickier.

2.2.   This is shown in Figure 6.1.

2.3. Γ_0 is a figure-8, and σ maps each circle to a path that traces both circles. $\tilde{\Gamma}_0$ has two vertices of valence 4. Two edges connect these directly (one in each direction), there is a path of length 2 from one to the other (via an intermediate vertex of valence 2), and there is a path back of length 2.

2.4.   This is described in detail in chapter 6.

Chapter 3.

3.2.   Let $\left(\begin{smallmatrix} n'_1 \\ n'_2 \end{smallmatrix}\right) = M\left(\begin{smallmatrix} n_1 \\ n_2 \end{smallmatrix}\right)$. Then $n'_1 + \tau n'_2 = n_2 + \tau(n_1 + n_2) = \tau(n_1 + \tau n_2)$. Put another way, when we decompose $\left(\begin{smallmatrix} n_1 \\ n_2 \end{smallmatrix}\right)$ into right-eigenvectors of M, the coefficient of the Perron-Frobenius eigenvector $R = \left(\begin{smallmatrix} 1 \\ \tau \end{smallmatrix}\right)$ is $(n_1 + \tau n_2)/(2 + \tau)$. Multiplying by M just scales this coefficient by τ.

3.3.   See the solution to the previous problem. $\left(\begin{smallmatrix} n_1 \\ n_2 \end{smallmatrix}\right)$ has a nonzero coefficient of the Perron-Frobenius eigenvector $R = \left(\begin{smallmatrix} 1 \\ \tau \end{smallmatrix}\right)$, which is positive if an only if $\left(\begin{smallmatrix} n_1 \\ n_2 \end{smallmatrix}\right)$ is positive. M^n enlarges that coefficient by τ^n, while shrinking the coefficient of $\left(\begin{smallmatrix} \tau \\ -1 \end{smallmatrix}\right)$ by $(1-\tau)^n$, so that $M^n \left(\begin{smallmatrix} n_1 \\ n_2 \end{smallmatrix}\right)$ asymptotically becomes a multiple of $\left(\begin{smallmatrix} 1 \\ \tau \end{smallmatrix}\right)$, with both n_1 and n_2 positive if the original coefficient is positive.

3.4.   Recall that the preimage of an intersection is the intersection of the preimages. The set $\mathcal{V}_i \cap \cdots \cap \mathcal{V}_p = \pi^{-1}(\mathcal{U}_i \cap \cdots \cap \mathcal{U}_p)$ is non-empty if and only if $\mathcal{U}_i \cap \cdots \cap \mathcal{U}_p$ is non-empty.

3.8.   The uncollared complex has $H^1 = \mathbb{Z}^2$ and $\sigma^* = \left(\begin{smallmatrix} 1 & 1 \\ 1 & 1 \end{smallmatrix}\right)$, which has rank 1. The direct limit is $\mathbb{Z}[1/2]$.

3.9. The collared complex has $H^1 = \mathbb{Z}^3$. The matrix for σ^* depends on your choice of basis for H^1, but has eigenvalues 2, 0, and -1. The direct limit is $\mathbb{Z} + \mathbb{Z}[1/2]$.

Chapter 4.

4.1. Since there are only a finite number of tile types, there are only a finite number of possible 1-tile patches (up to Euclidean motion). For each of these, there is only a finite number of ways to add a second tile to get a 2-tile patch, insofar as the new tile has to have an edge that fits exactly with an existing edge. By induction, there are only finitely many n-tile patches, which immediately implies FLC.

4.2. If two collared tiles play the same role at a vertex, then the pattern around the vertex must by invariant under a rotation that takes one tile to the other, and hence by the group generated by that rotation. Since there are only a finite number of tiles meeting at that vertex, that group must be finite, so the rotation must be by a rational multiple of π.

Chapter 5.

5.1. Let ω be a strongly PE form with radius R, and let $R' > R$. Suppose x and y are points whose neighborhoods of size R' agree. Then ω takes on the same values on a ball of radius $R' - R$ around x as on a similar ball around y, and so the partial derivatives of ω at x agree with those at y. Thus $d\omega$ is strongly PE with radius R'.

5.2. Suppose f is a PE function. Since f and its first k derivatives are well-approximated by strongly PE forms, df and its first $k - 1$ derivatives are well-approximated by strongly PE 1-forms. Since both k and the degree of approximation are arbitrary, df is weakly PE. The same argument applies to forms of higher degree.

5.3. Let x be the position of the left endpoint of the black tile. A strongly PE function is a function of x that is periodic with period 1 for $|x|$ large. $f(x) = \sin(x^4)/(1 + x^2)$ is the uniform limit of such functions, since it is approximated well by the periodic function 0 for x large. However, $f'(x)$ behaves wildly near infinity and is not weakly PE.

5.4. For each collared tile t, there is an ϵ_t such that the ring of tiles around t contains an ϵ_t-neighborhood of t, and let ϵ the the smallest such ϵ_t. Let D be the largest diameter of any tile. Then the ring of tiles around any fixed tile contains an ϵ-neighborhood of that tile and is contained in a D-neighborhood of that tile. Let $r_n = n\epsilon$ and let $R_n = nD$.

5.5. All points in a $k + 1$-cell c are within D of the center of mass. In particular, a ball of radius $R + D$ around the center of mass of c contains balls of radius R around the center of mass of all components of ∂c. If c and c' agree to radius $R + D$, then their boundaries agree to radius r, and $(\delta\alpha)(c) = \alpha(\partial c) = \alpha(\partial c') = (\delta\alpha)(c')$.

5.7. If $[\alpha] = [\alpha']$, then $\alpha - \alpha' = \delta\beta$, for some PE cochain β. If c is a large chain whose boundary is small compared to its area, $\alpha(c) - \alpha'(c) = \beta(\partial c)$, so the average of α over c is close to the average of α'. Taking the limit as c grows gives identical averages over the whole tiling.

Chapter 6.

6.1. A substitution forces the border if some power of the substitution forces the border, and proper substitutions force the border.

6.2. Look at the vertices that appear as boundaries of supertiles of high order.$n > N$, where the substitution forces the border in N steps. Since an "a" supertile has a definite end and the subsequent supertile has a definite beginning, and likewise for a "b" supertile, there are at most two possible vertices that separate supertiles of order n. If there is only one, then σ^n is proper. Suppose that there are two vertices, $a.a$ and $a.b$. Then every supertile of order n ends in a, and every supertile of order $2n$ ends in a and is followed by whatever letter always follows $\sigma^n(a)$. This implies that σ^{2n} is proper. Similar arguments apply to all pairs except $\{a.a, b.b\}$ and $\{a.b, b.a\}$. But in those cases, the sequence of order n supertiles must be periodic, since each supertile determines its successor.

6.3. Every letter of the new alphabet begins with y and ends with x. After n substitutions, every supertile begins with $\sigma^n(y)$ and ends with $\sigma^n(x)$. If n is large enough, $\sigma^n(x)$ ends with a complete letter of the new alphabet, and $\sigma^n(y)$ begins with a complete letter of the new alphabet, making σ^n proper.

6.4. $A = (a)ABBA(a)$, $B = (a)ababba(a)$, $C = (a)abbaba(a)$, $D = (a)ababbaba(a)$, $\sigma(A) = CAB$, $\sigma(B) = CDAB$, $\sigma(C) = CADB$, and $\sigma(D) = CDADB$.

6.5. After a fixed number of substitutions, a partially collared tile determines its immediate neighbors as ordinary tiles. After several more substitutions, these neighbors determine a region at least two tiles thick around the substituted collared tile, thereby determining the once-collared tiles (and hence the partially-collared tiles) that touch the substituted partially collared tile.

6.6. $A \to \left(\begin{smallmatrix} r^3F & A \\ B & rD \end{smallmatrix}\right)$, $B \to \left(\begin{smallmatrix} r^3F & B \\ B & rD \end{smallmatrix}\right)$, $C \to \left(\begin{smallmatrix} r^3F & C \\ B & rD \end{smallmatrix}\right)$, $D \to \left(\begin{smallmatrix} r^3F & A \\ E & rD \end{smallmatrix}\right)$, $E \to \left(\begin{smallmatrix} r^3F & B \\ E & rD \end{smallmatrix}\right)$, $F \to \left(\begin{smallmatrix} r^3F & C \\ E & rD \end{smallmatrix}\right)$. The order-2 supertiles are all of the form
$$\begin{pmatrix} r^3E & r^2F & r^3F & A/B/C \\ D & r^3C & B & rD \\ r^3F & B & rA & r^2D \\ B/E & rD & F & rE \end{pmatrix}.$$

6.7. The identifications are shown below. The rotation group $\mathbb{Z}_4$ acts trivially on the vertices a, c and d and freely on the vertex b and the edges and faces. In the trivial representation we have 6 faces, 6 edges and 4 vertices. In the other representations we have 6 faces, 6 edges and one vertex.

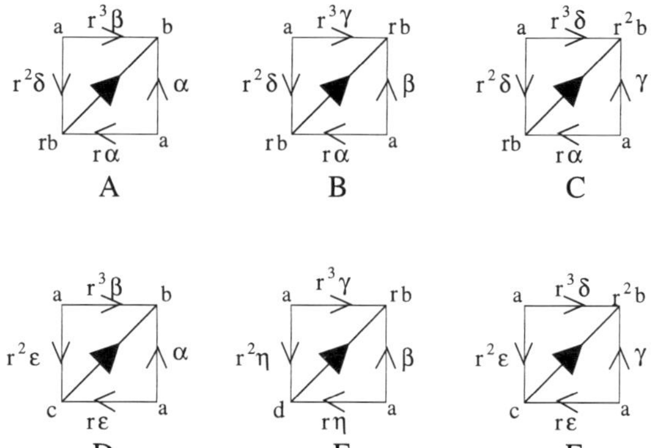

6.8. The matrices are:

$$\partial_1 = \begin{pmatrix} 1 & r & r^2 & r^3 & 0 & 0 \end{pmatrix} \qquad \text{if } r \neq 1,$$

$$\partial_1 = \begin{pmatrix} -1 & -1 & -1 & -1 & -1 & -1 \\ 1 & 1 & 1 & 1 & 0 & 0 \\ 0 & 0 & 0 & 0 & 1 & 0 \\ 0 & 0 & 0 & 0 & 0 & 1 \end{pmatrix} \qquad \text{if } r = 1$$

$$\partial_2 = \begin{pmatrix} 1-r & -r & -r & 1 & 0 & 0 \\ -r^3 & 1 & 0 & -r^3 & 1 & 0 \\ 0 & -r^3 & 1 & 0 & -r^3 & 1 \\ r^2 & r^2 & r^2 - r^3 & 0 & 0 & -r^3 \\ 0 & 0 & 0 & r^2 - r & 0 & r^2 - r \\ 0 & 0 & 0 & 0 & r^2 - r & 0 \end{pmatrix}. \qquad \text{(A.1)}$$

The matrices of δ_0 and δ_1 are the transposes of ∂_1 and ∂_2 with r replaced by r^3.

6.9. In the $r = 1$ representation, δ_0 has rank 3 and δ_1 has rank 3, with kernel equal to the image of δ_0. We thus have $H^0 = \mathbb{Z}$ with generator $(1,1,1,1)^T$, $H^1 = 0$, and $H^2 = \mathbb{Z}^3$. The six generators of H^2 are the duals $A', \ldots, F'$ to the six tiles, and have the three relations $D' = B' + C'$, $E' = A' + C'$, $F' = A' + B'$. In the $r = -1$ representation, δ_1 has rank 5, with kernel equal to the image of δ_0. $H^2 = \mathbb{Z} \oplus \mathbb{Z}_2 \oplus \mathbb{Z}_4$, with six generators and five relations given by the columns of δ_1. In the 2-dimensional $r^2 = -1$ representation, δ_1 has rank 4. H^1 is one copy of the representation (i.e., $H^1 = \mathbb{Z}^2$), with generator $(1,1,1,1,1,1)^T$, and H^2 has two copies, expressed as six generators with relations $A' = -C'$, $B' = r^3 C' + r^3 F'$, $D' = -F'$, $E' = 0$.

6.10. The substitution matrix on 2-chains and the induced matrix on 2-cochains are

$$
\sigma_2 = \begin{pmatrix}
1 & 0 & 0 & 1 & 0 & 0 \\
1 & 2 & 1 & 0 & 1 & 0 \\
0 & 0 & 1 & 0 & 0 & 1 \\
r & r & r & r & r & r \\
0 & 0 & 0 & 1 & 1 & 1 \\
r^3 & r^3 & r^3 & r^3 & r^3 & r^3
\end{pmatrix}
$$

$$
\sigma_2^* = \begin{pmatrix}
1 & 1 & 0 & r^3 & 0 & r \\
0 & 2 & 0 & r^3 & 0 & r \\
0 & 1 & 1 & r^3 & 0 & r \\
1 & 0 & 0 & r^3 & 1 & r \\
0 & 1 & 0 & r^3 & 1 & r \\
0 & 0 & 1 & r^3 & 1 & r
\end{pmatrix},
\tag{A.2}
$$

while the matrices on 1-chains and cochains are

$$
\sigma_1 = \begin{pmatrix}
1 & 0 & 0 & 0 & 0 & 0 \\
0 & 1 & 0 & 0 & 0 & 0 \\
0 & 0 & 1 & 0 & 0 & 0 \\
0 & 0 & 0 & 1 & 0 & 0 \\
-r^2 & -r^2 & -r^2 & -r^2 & -r^2 & -r^2 \\
0 & 0 & 0 & 0 & 1 & 1
\end{pmatrix},
$$

$$
\sigma_1^* = \begin{pmatrix}
1 & 0 & 0 & 0 & -r^2 & 0 \\
0 & 1 & 0 & 0 & -r^2 & 0 \\
0 & 0 & 1 & 0 & -r^2 & 0 \\
0 & 0 & 0 & 1 & -r^2 & 0 \\
0 & 0 & 0 & 0 & -r^2 & 1 \\
0 & 0 & 0 & 0 & -r^2 & 1
\end{pmatrix}.
\tag{A.3}
$$

The substitution on vertices sends a to c to d, and leaves b and d alone.

In the $r = 1$ representation, we can take the generators of H^2 to be A', B' and C'. Substitution sends these to cochains that are equivalent to $A' + B' + C'$, $2(A' + B' + C')$, and $A' + B' + C'$, respectively. That is, the direct limit of H^2 is generated by $A' + B' + C'$, which is multiplied by 4 on rotation. The limit is therefore $\mathbb{Z}[1/4]$.

In the $r = -1$ representation, the columns of $(\sigma^*)^2$ are all integer linear combinations of the columns of δ_1, so that all terms in H^2 are killed by repeated substitution, and the direct limit is zero.

In the $r^2 = -1$ representation, the generator of H^1 is $(1, 1, 1, 1, 1, 1)^T$, which simply counts edges regardless of type. This gets multiplied by 2 under substitution, so our direct limit is $\mathbb{Z}[1/2]^2$. As for H^2, the generators C' and F' each get mapped to cochains cohomologous to $C' + F'$, so $\check{H}^2$ is $\mathbb{Z}^2$ tensored with the direct limit of $\left(\begin{smallmatrix} 1 & 1 \\ 1 & 1 \end{smallmatrix}\right)$, that is $\mathbb{Z}[1/2]^2$.

The upshot is that $\check{H}^0 = \mathbb{Z}$, $\check{H}^1 = \mathbb{Z}[1/2]^2$, coming from the $r^2 = -1$ representation, and $\check{H}^2 = \mathbb{Z}[1/4] \oplus \mathbb{Z}[1/2]^2$, with the $\mathbb{Z}[1/4]$ coming from the $r = 1$ representation and the $\mathbb{Z}[1/2]^2$ coming from the $r^2 = -1$ representation.

6.11. We have already seen that $(1,1,1,1,1,1)^T$, which doesn't distinguish between collared tiles, is the the generator of $\check{H}^1$. The generator of the $\mathbb{Z}[1/2]$ term in $\check{H}^2$ is $C' + F'$, which in this representation is cohomologous to $A' + B' + C' + D' + E' + F'$, which counts tiles independent of collaring. However, the generator of the $\mathbb{Z}[1/4]$ term is $A' + B' + C'$, which is a third of $A' + B' + C' + D' + E' + F'$. That is, the generator of the "round earth" $Z[1/4]$ is a third of the generator of the "flat earth" $\mathbb{Z}[1/4]$.

6.13. The answer depends on the basis we pick for $H_1(S_0)$. If we use the basis $\{b-c-f+h, -b+c-e+f-j+l, a+f+i+n, a+b-c+e+j+p\}$, obtained by row-reducing ∂_1 and column-reducing ∂_2, then $\delta = \begin{pmatrix} 1 & 1 & 0 & 1 \\ -1 & 0 & 1 & 0 \\ 0 & -1 & 0 & 0 \\ 0 & 0 & 2 & 1 \end{pmatrix}$, which has determinant -3.

Chapter 7.

7.1. The small triangle has sides of length 1, $\sqrt{\frac{\sqrt{17}-1}{2}}$ and $\sqrt{\frac{\sqrt{17}+1}{2}}$. The big triangle has sides of $\sqrt{\frac{\sqrt{17}+1}{2}}$, 2, and $\frac{\sqrt{17}+1}{2}$.

7.3. The Mayer-Vietoris sequence reads

$$
\begin{aligned}
0 \to \tilde{H}^0(S_\Sigma/S_1) &\to 0 \to \mathbb{Z} \\
\hookrightarrow \tilde{H}^1(S_\Sigma/S_1) &\to H^1(S_\sigma) \to H^1(S_\sigma) \oplus H^1(S_\sigma) \\
\hookrightarrow \tilde{H}^2(S_\Sigma/S_1) &\to 0 \to 0.
\end{aligned}
\tag{A.4}
$$

Since each component of $U \cap V$ is a deformation retract of U, the last map on the second row (call it δ) is injective, and $\delta(x) = (x, -x)$, so $H^1(S_\Sigma/S^1) = \tilde{H}^1(S_\Sigma/S^1) = \mathbb{Z}$ and $H^2(S_\Sigma/S^1) = \tilde{H}^2(S_\Sigma/S^1) = H^1(S_\sigma)$.

7.4. Since x can follow x, X can follow x, x can follow X but X cannot follow X, this is the same as the Barge-Diamond complex for the Fibonacci tiling, only with X and x playing the roles of a and b. $S_{\rho,0}$ is contractible, so $H^0 = \mathbb{Z}$ and $H^1 = 0$.

7.5. By corollary 6.6, H^1 is the direct limit of the transpose of the substitution matrix, i.e., $\mathbb{Z} \oplus \mathbb{Z}[1/2]$.

7.6.   Since U retracts to $S_\sigma * S_\sigma$, as does V, while $U \cap V$ retracts to S_σ, our Mayer-Vietoris sequence reads

$$
\begin{aligned}
0 \to H^1((S_1)_{\mathrm{ER}}) \quad &\to \qquad\qquad 0 \qquad\qquad \to \quad H^1(S_\sigma) \\
\hookrightarrow \quad H^2((S_1)_{\mathrm{ER}}) \quad &\to \qquad\qquad 0 \qquad\qquad \to \qquad 0 \\
\hookrightarrow \quad H^3((S_1)_{\mathrm{ER}}) \quad &\to \quad [H^1(S_\sigma) \otimes H^1(S_\sigma)]^2 \quad \to \qquad 0.
\end{aligned}
\tag{A.5}
$$

which immediately determines $H^*((S_1)_{\mathrm{ER}})$.

7.7.   In our decomposition of $(S_1)_{\mathrm{ER}}$, U and V are joined at the bottom of a supertile, and $H^2((S_1)_{\mathrm{ER}})$ comes from the $H^1(S_\sigma)$ term in $H^1(U \cap V)$, so Σ^* acts on $H^2((S)_{\mathrm{ER}})$ via the substitution that applies at the bottom of a supertile, namely σ_1. However, the two $[H^1(S_\sigma) \otimes H^1(S_\sigma)]$ terms in $H^2((S_1)_{\mathrm{ER}})$ come from $H^3(U)$ and $H^3(V)$, each of which is the join of two pieces, one that substitutes via σ_1 and one that substitutes via σ_2.

Bibliography

[AP]   J. E. Anderson and I. F. Putnam, Topological invariants for substitution tilings and their associated C^*−algebras, *Ergodic Theory Dynam. Systems* **18** (1998), 509–537.

[BBG]   J. Bellissard, R. Benedetti, and J.-M. Gambaudo, Spaces of tilings, finite telescopic approximations and gap-labelling, *Comm. Math. Phys.* **261** (2006) 1–41.

[BD]   M. Barge and B. Diamond, Cohomology in one-dimensional substitution tiling spaces, *Proc. Amer. Math. Soc* **138** (2008) 2183–2191.

[BDHS]   M. Barge, B. Diamond, J. Hunton and L. Sadun, in preparation.

[BG]   R. Benedetti, J.-M. Gambaudo, On the dynamics of G-solenoids: applications to Delone sets, *Ergodic Theory and Dynamical Systems* **23** (2003) 673–691.

[Ber]   R. Berger, The undecidability of the Domino problem. *Mem. Amer. Math. Soc.* **66** (1966).

[BSJ]   M. Baake, M. Scholottmann and P.D. Jarvis, Quasiperiodic tilings with tenfold symmetry and equivalence with respect to local derivability, *J. Phys. A.* **24** (1991) 4637–4654.

[BT]   R. Bott and L. Tu, "Differential Forms in Algebraic Topology", Springer-Verlag, 1982.

[Cul]   K. Culik II, An aperiodic set of 13 Wang tiles, *Discrete Math.* **160**, 245–251.

[CS]   A. Clark and L. Sadun, When Shape Matters: Deformations of Tiling Spaces, *Ergodic Theory and Dynamical Systems* **26** (2006) 69–86.

[Da]   Inflation species of planar tilings which are not of locally finite complexity. *Proc. Steklov Inst. Math.* **230** (2002), 118–126.

[Dur]   A characterization of substitutive sequences using return words. *Discrete Math.* **179** (1998), 89–101.

[FHK]   A. Forrest, J. Hunton and J. Kellendonk, Topological Invariants for Projection Method Patterns, Memoirs of the Amer. Math. Soc. **758**, (2002).

[FR]   N.P. Frank and E.A. Robinson, Jr., Generalized β-expansions, substitution tilings, and local finiteness, *Trans. Amer. Math. Soc.* **360** (2008) 1163–1177.

[Fr]   N.P. Frank, A primer on substitution tilings of Euclidean space, to appear in *Expositiones Mathematicae.*

[FS]   N.P. Frank and L. Sadun, Topology of (Some) Tiling Spaces without Finite Local Complexity, to appear in *Disc. & Cont. Dyn. Sys.*

[Gah]   F. Gähler, talk given at the conference "Aperiodic Order, Dynamical Systems, Operator Algebras and Topology", 2002, slides available at `www.pims.math.ca/science/2002/adot/lectnotes/Gaehler`

[GS]   C. Goodman-Strauss, Matching Rules and Substitution Tilings, *Annals of Mathematics*, **147** (1998), 181–223.

[Hat]   A. Hatcher "Algebraic Topology", (2002) Cambridge University Press.

[Kar]   J. Kari, A small aperiodic set of Wang tiles, *Discrete Math.* **160**, 259–264.

[Kel1]   Noncommutative geometry of tilings and gap-labelling, *Rev. Math. Phys.* **7** (1995), , 1133–1180.

[Kel2]   J. Kellendonk, Pattern-equivariant functions and cohomology, *J. Phys. A* **36** (2003) 5765–5772.

[Ken]   R. Kenyon, Self-Replicating Tilings, in "Symbolic dynamics and its applica-
          tions", AMS Contemp. Math. Series 135, P. Walters ed., (1992) 239–263.

[KP]   J. Kellendonk and I. Putnam, The Ruelle-Sullivan map for R^n actions, *Math.
          Ann.* **334** (2006), 693–711.

[Moo]   R. Moody, Model sets: A Survey, in "From Quasicrystals to More Complex Sys-
          tems", eds. F. Axel, F. Dénoyer, J.P. Gazeau, Centre de physique Les Houches,
          Springer Verlag, 2000.

[Mos]   B. Mossé: Puissances de mots et reconnaissabilité des point fixes d'une substi-
          tution. *Theor. Comput. Sci.* **99** (1992) 327–334.

[Moz]   S. Mozes, Tilings, substitution systems and dynamical systems generated by
          them, *J. Analyse Math.* **53** (1989), 139–186.

[ORS]   N. Ormes C. Radin and L. Sadun, A Homeomorphism Invariant for Substitution
          Tiling Spaces, *Geometriae Dedicata* **90** (2002), 153–182.

[Pen]   M. Gardner, Extraordinary nonperiodic tiling that enriches the theory of tiles,
          Scientific American, December 1977, 110–119.

[Pet]   K. Petersen, Factor maps between tiling dynamical systems *Forum Math.* **11**
          (1999), 503–512.

[PS]   N. Priebe and B. Solomyak, Characterization of Planar Pseudo-Self-Similar
          Tilings, *Disc. & Comp. Geom.* **26** (2001) 289–306.

[Rad]   C. Radin, The pinwheel tilings of the plane, *Annals of Math.* **139** (1994) 661–702.

[Ran]   B. Rand, Pattern Equivariant Cohomology of Tiling Spaces with Rotations, PhD
          thesis in mathematics (2006) University of Texas

[Ran2]   B. Rand, Equivalence of Self-Similar and Pseudo-Self-Similar Tilings Spaces in
          $\mathbb{R}^2$, submitted.

[Rob]   R.M. Robinson, Undecidability and nonperiodicity for tilings of the plane, *In-
          vent. Math.* **12** (1971) 177–209.

[RS]   C. Radin and L. Sadun, Isomorphism of Hierarchical Structures, *Ergodic Theory
          Dynam. Systems* **21** (2001), no. 4, 1239-1248.

[Sa1]   L. Sadun, Some Generalizations of the Pinwheel Tiling, *Disc. Comp. Geom.* **20**
          (1998), 79–110.

[Sa2]   L. Sadun, Tiling spaces are inverse limits, *J. Math. Phys.* **44** (2003) 5410–5414.

[Sa3]   L. Sadun, Pattern Equivariant Cohomology with Integer Coefficients, *Ergodic
          Theory Dynan. Sys.* **27** (2007) 1991–1998.

[SBGC]   D. Shechtman, I. Blech, D. Gratias, and J. W. Cahn, Metallic Phase with Long-
          Range Orientational Order and No Translational Symmetry, *Phys. Rev. Lett.* **53**,
          (1984) 1951–1953.

[Sol]   B. Solomyak, Nonperiodicity implies unique decomposition for self-similar trans-
          lationally finite tilings. *Disc. & Comp. Geom.* **20** (1998), 265–279.

[SW]   L. Sadun and R.F. Williams, Tiling Spaces are Cantor Set Fiber Bundles, *Ergodic
          Theory and Dynamical Systems* **23** (2003) 307–316.